Cannabis Use in Medici

Rahim Valani

Editor

Cannabis Use in Medicine

A Concise Handbook

 Springer

Editor
Rahim Valani
Department of Medicine
University of Toronto
Toronto, ON, Canada

ISBN 978-3-031-12721-2   ISBN 978-3-031-12722-9   (eBook)
https://doi.org/10.1007/978-3-031-12722-9

This Springer imprint is published by the registered company Springer Nature Switzerland AG
The registered company address is: Gewerbestrasse 11, 6330 Cham, Switzerland

This book is dedicated to justice, healing, and reconciliation with the Red River Métis community and all Indigenous Peoples of Canada.

This acknowledgement serves as a reminder to us all to continually renew our efforts to recognize, honor, reconcile, and partner to celebrate the strength of the Red River Métis and all Indigenous People across Canada and the world.

Preface

Cannabis use for symptom management is relatively new in the clinical realm. While cannabis has been used for centuries, the medical community is now beginning to understand how this product can be used. Recognizing the increased use of cannabis, clinicians must know the uses as well as understand the complications in relation to cannabis use. As clinicians, we are always looking to expand our knowledge and skills, and to use the best evidence to ensure excellent care for our patients. *Cannabis Use in Medicine: A Concise Handbook* brings together the knowledge and expertise of clinicians in a succinct form for easy reference.

This handbook is divided into three parts. The chapters in Part I provide background on cannabis that covers legal aspects, pharmacology, genetics, and patient assessment. Part II focuses on specific systems where cannabis has shown benefit. Finally, Part III deals with specific populations.

Despite the ongoing growth of knowledge and experience in cannabis use, there are currently few reference materials available. This handbook will help provide a primer on this topic. One book, however, cannot cover everything, and for this reason this book is designed as a concise handbook.

As with any reference source, there are always opportunities to improve. I invite feedback so that this handbook can continue to evolve as a guide for clinicians who work with patients who are using cannabis, as well as seasoned practitioners who prescribe cannabis for medical conditions who wish to share their wealth of expertise.

Toronto, ON, Canada   Rahim Valani

Acknowledgments

This book would not have been possible without the vision and dedication of several people. I would like to express my appreciation to all the contributing chapter authors for the time and expertise they devoted to putting this book together, on top of their already demanding schedules—thank you for your excellent work.

I am grateful to those who encouraged me to embark on this project so that we could develop a resource for clinicians. My sincere thanks to Wade Attwood, Michael Dunkerly, Salil Dhaumya, Dave Coolidge, Marisa Cornacchia, and Meghan McKenna for their tremendous support. I would also like to acknowledge Dr. David Chartrand and the Manitoba Metis Federation for their support.

To my colleagues at work, I am appreciative of your patience with me during the countless hours I spent completing this book. In addition, I want to thank the team at Springer for their professional guidance and support in bringing this book to fruition.

Contents

Introduction and Background to Cannabis Use in Medicine

Introduction to Cannabis for Medical Use

1

Rahim Valani

Cannabis is one of the most used recreational substances worldwide. It has been used for a variety of other reasons as well, including medical and industrial purposes. With legalization of cannabis for recreational and medical use, the clinician needs to better understand the indications as well as adverse effects of cannabis use. Further research is shedding more light on the benefits of cannabis for symptoms such as chronic pain and intractable seizures.

Introduction

- Cannabis is one of the most commonly used substances worldwide. It has been used for centuries for recreational, medical, and industrial use.
  - While the term "medical cannabis" has been used routinely, this phrase should be discouraged. It implies a solitary product or one that is fully endorsed by the medical community.

    Cannabis comes in a variety of strains with different concentrations of phytocannabinoids.

    Not all clinicians are comfortable prescribing cannabis.
  - Regional use of cannabis is highest in North America, Oceania, and West Africa [1].
  - Adult males are the largest consumers of recreational cannabis.
- Cannabis has been known by names, the most common being marijuana, weed, pot, ganja, and Mary Jane. The primary product is the dried flowers of *Cannabis Sativa* plant.

R. Valani (✉)
Department of Medicine, University of Toronto, Toronto, ON, Canada
e-mail: r.valani@utoronto.ca

© Springer Nature Switzerland AG 2022
R. Valani (ed.), *Cannabis Use in Medicine*,
https://doi.org/10.1007/978-3-031-12722-9_1

- Cannabis has been used in various forms, the most common being:
  - Smoking and vaping (pulmonary route).
  - Edibles, including tea and other food products (gastrointestinal route).
  - Creams and ointments (dermal route).
- The majority of users do not experience any severe effects. Those who use high doses and on a daily basis are more prone to serious adverse effects.
  - The high lethal dose (estimated at over 15 g of THC) is well above the recommended dose. Furthermore, it does not cause respiratory depression like opioids making it a great alternative or adjunct to opioids [2].
- Cannabis for medical conditions is being actively researched. The two main compounds, Δ9-Tetrahydrocannabinol (THC) and Cannabidiol (CBD), are the primary focus for much of the research.
- Cannabis cultivators classify the type of plant/product under the following chemotypes [3]:
  - Chemotype I—The THC:CBD ratio is $\geq$10.
  - Chemotype II—the THC:CBD ratio is between 0.2 and 10.
  - Chemotype II is usually seen with fibre type cultivators, and this contains THC:CBD <0.2.
  - Chemotype IV is predominantly Cannabigerol.
- There are challenges for using cannabis for medical purposes as well as developing the right type of clinical trials. Much of this stems from the legal issues as well as the stigma associated with cannabis use. Some of the challenges in building the evidence for cannabis use include [4]:
  - No single source of cannabis makes it difficult to compare strains, concentration, and quality of the substance.
  - Difficulty in designing the clinical trials (randomization, source of cannabis, security measures, etc.).
  - Approval to use cannabis for research (e.g., in the US, it would require FDA approval, registration with the Drug Enforcement Agency, and approval from the research ethics board).
  - Finding agencies that will fund the study.

Legalization of cannabis use see Chap. 2.

- The first documented international restriction for cannabis was the International Convention of 1928 which prohibited recreational use [5].
- In 1937, the US introduced the Marijuana Tax Act which made it expensive to deal with cannabis in any capacity. By 1941, it was removed from the US national formulary.
  - California was the first state in the US to legalize cannabis for medical use.
- Canada passed the Act to Prohibit the Improper Use of Opium and other Drugs in 1923, and cannabis was listed as one of the prohibited drugs. It is not until 2018 that the Cannabis Act was passed which legalized the consumption and sale of cannabis for recreational and medicinal purposes.

- Legalization cannabis use for non-medical purposes has been a challenge. Many countries have introduced laws that provide legal access to cannabis for medical conditions. The role of physicians in prescribing or facilitating access to cannabis for treatment varies by jurisdiction. Some of the countries that have developed laws for access to medical cannabis include:
  - Uruguay—2013.
  - Canada—2018.
  - Mexico—2021.
  - USA.
  - United Kingdom.
  - Australia.
  - Germany.
- Having these laws benefit patients by:
  - Providing control over the quality of cannabis produced and dispensed.
  - Ensure appropriate measures are in place for processing extracts in order to avoid impurities or toxic substances.
  - It creates appropriate distribution channels for the end user, thus avoiding illegal sales and purchases.
  - Having medical oversight of patients who use cannabis, and therefore appropriately dose patients and reduce complications/side effects.

The endocannabinoid system see Chap. 3.

- The endocannabinoid system consists of two primary receptors, namely CB_1 and CB_2 [6].
  - CB_1 receptors are seen throughout the nervous system.
    It is a presynaptic receptor that is involved in neurotransmitter release/inhibition.
    Effects of this receptor include:
    - Euphoric effects.
    - Hypotension.
    - Anti-inflammatory action.
    - Immunosuppression.
    - Analgesic activity.
    - Appetite stimulant.
  - CB_2 receptors are found in tissues of the immune system, liver, and some neurons.
    CB_2 receptors, on the other hand, are associated with anti-inflammatory effects and immune modulation.
- There are native ligands to these receptors that help in regulating homeostasis. The phytocannabinoids, such as THC and CBD, have different affinities for these receptors resulting in different physiological effects.
- The fist step in cannabinoid biosynthesis is the production of olivetoic acid. However, the synthesis and production of cannabinoids is not well elucidated.

Table 1.1 Cannabinoids and their effects

Compound	Receptor affinity	Potential effects
Cannabigerol	Low affinity for CB_1 and CB_2	Antineoplastic
Cannabichromen		Antidepressant
Cannabidiol	Weak antagonist of CB_1 and CB_2	Analgesia, anti-inflammatory, anxiolysis
Δ9-tetrahydrocannabivarin	Partial CB_2 agonist and CB_1 antagonist	Antiepileptic
Δ9-tetrahydrocannabinol	CB_1 agonist	Appetite stimulant, helps with sleep
Δ8-tetrahydrocannabinol	CB_1 agonist	Anti-glaucoma

Analytical techniques are discovering new compounds from the plant and the effects are being studied. The effects of THC and CBD have the longest track record and continue to be the two most common cannabinoids studied. See Chap. 2.

- The effects of other cannabinoids are summarized in Table 1.1 [7].

Recreational Cannabis Use

- Cannabis for recreational purposes has been around for ages. While low doses of cannabis use have been shown to be safe, the best form of harm reduction would be to abstain from cannabis use.
- The international guidelines for lower-risk cannabis use highlight several recommendations for recreational cannabis use, some of which include [1]:
  - Avoiding cannabis use until after adolescence. The reason cited is to reduce any developmental related risk.
  - There are many synthetic and novel strains of cannabis being produced. The recommendation is to use products with lower concentrations of THC so as to avoid its effects.
  - Avoid frequent (daily) use or binge use.
  - Quality and safety is important with any drug or compound use. Ensure use of legal and quality-controlled cannabis products.
  - If there are any signs of cognitive impairments with use, then the person should stop using cannabis (or at least reduce consumption).
  - Avoid safety sensitive tasks while on cannabis.
  - Pregnant women and new mother should abstain from cannabis use.
- Synthetic cannabinoids are used for recreational use given the higher potency of THC compared to the native cannabis plant. Adverse effects of these synthetic compounds are, not surprisingly, intensified. These effects are summarized in Table 1.2.
- There has also been a shift in the type of cannabis chemotype being cultivated, with increasing THC concentrations being seen. THC concentration in cannabis has been increasing, and more adverse effects are observed.

Table 1.2 Adverse effects of synthetic cannabinoids [9]

System	Effects
Psychiatric	Severe psychotic symptoms
	Agitation
	Paranoia
	Altered perception/hallucinations
	Negative mood effects/depression
	Panic attacks
	Suicidal ideation
Neurological	Cognitive impairment
	Memory alteration/lapses
Cardiovascular	Tachycardia
	Hypertension
	Arrhythmias
Neurological	Dizziness
	Hypertonicity, hyper-reflexia
	Altered sensory perception
Gastrointestinal	Nausea/vomiting
	Abdominal pain

- Over the past decade, THC concentration has been steadily climbing based on confiscated cannabis [8]:
- In 2009, cannabis confiscated had a THC concentration of 9.8%.
- For 2014, this had increased to 11.7%.
- By 2019, it was 13.9%.

The Use of Cannabis for Medical Conditions

- Patients have been using cannabis for self treatment of symptoms. While patients describe relief of symptoms and better quality of life, these results are not well studied in the medical literature.
  - From a mental health perspective, patients report using cannabis for anxiety and depression with positive effects [10]. However, clinical studies have not shown the same benefits.
- The medical community has been reticent to embrace cannabis as a medicine for a multitude of reasons [11]:
  - Misinformation based on traditional models of medical education or societal perception of cannabis.
  - Poorly designed clinical studies which do not show a benefit of cannabis over traditional medical treatment.
  - Lack of a specific quality or standardized products, which makes comparison between patients a challenge. Furthermore, comparison between trials or trying to merge trials (meta-analysis) is a challenge for this reason.
  - The patients being studied are heterogeneous which may impact the results.

- Use of varying definitions can limit conclusions. For example, a study on pain may show no benefit. However, the study may mix patients with different types and severity of pain making hard to reach a positive conclusion.
- The indications for cannabis use continues to increase as we understand the mechanism of action and physiological effects. At present, cannabis has been shown to have beneficial effects for the following conditions [12–14]:
  - Chronic and cancer related pain. See Chap. 11.
  - Nausea and vomiting caused by chemotherapy.
  - Spasticity of multiple sclerosis. See Chap. 9.
  - Sleep disorders. See Chap. 9.
  - Glaucoma.
  - Refractory seizure disorders. See Chap. 9.
  - Nociplastic pain and inflammation related to rheumatological conditions such as fibromyalgia. See Chap. 10.
- Cannabis use can have a variety of effects, both in acute and chronic use. See Chap. 3.
  - In the acute stages of cannabis use, there are no changes to the cannabinoid receptors.
    Evidence exists for decreased serotonin effects, decreased blood flow to the cerebellum, and decreased cerebral metabolism [15].
  - In chronic users, there is downregulation of cannabinoid receptors. This may require additional doses to obtain the same desired effect.
- In patients using cannabis, a careful medication history is important to ensure no drug interaction.

Cannabis Products

- Cannabis can be administered in various forms depending on the comfort of the user, the condition it is being used for, and the dosage required. See Chap. 3. The most common routes of administration are:
  - Oral.
    This is the easiest way to monitor and dose.
    There are various formulations available that contain different amounts of THC and CBD.
    One of the down sides of this route is the lower bioavailability rate since the compounds are subject to first pass metabolism.
  - Inhalation pulmonary.
    Smoking or vaping are common inhalational modes of administration.
    While the system concentration is high, there are concerns related to the combustion that produce resultant toxins.
  - Dermal route.
    Creams and oils have been used for local application.
    Cannabis is not benign, and it depends on how it is being used [16].

- It effects on different population can vary such as children, pregnant, and elderly patients.
  - Elderly [17, 18]—cognitive effects; drug-drug interactions.
  - Pregnancy—effects on the mother and developing fetus. See Chap. 15.
- Apart from the cannabis plant, there are several pharmaceutical products available on the market. These can either be extracts from the plant or synthetic compounds developed to mimic the effects of either THC or CBD. See Table 1.3.
- The different products are approved for various indications.
  - Nabiximols (Sativex)

    This is a refined Cannabis extract product that is administered as an oromucosal spray.

    It contains a balanced THC:CBD product, where each puff administers 2.7 mg of THC and 2.5 mg of CBD.

    Approved for the use of spasticity in patients with multiple sclerosis.
  - Nabilone (Cesamet, Canemes)

    This is a synthetic cannabinoid that is administered as a capsule.

    It mimics the effects of THC and primarily works on CB_1 receptors.

    This medication is used for the treatment of nausea and vomiting due to chemotherapy.
  - Dronabilone (Marinol, Syndros, Reduvo, Adversa)

    This product is a synthetic product of THC that is available as a capsule.

    It is recommended to be used as an appetite stimulant (AIDS related anorexia) and chemotherapy associated emesis.
  - Epidiolex

    This product is purified CBD that is administered as an oral liquid.

    This products is indicated for the treatment of seizures (Lennox-Gastaut syndrome, Dravet syndrome, and tuberous sclerosis) for patients over the age of 1 year.
  - Other products in development:

    Extracted THC (Namisol)

    Extracted CBD (Arvisol)

Table 1.3 Pharmaceutical cannabis products

	THC dominant	Balanced THC:CBD	CBD dominant
Extract	Namisol	Nabiximols	Epidiolex Arvisol
Synthetic	Nabilone Dronabiolone		

The Use of Cannabis in Oncology: A Case Example

- Cannabis has been used by cancer patients for both symptom treatment and attempts at curing the cancer. While data on the latter are limited in the medical literature, symptom management is well established.
  - Chemotherapy associated nausea and vomiting [19].
    Nausea and vomiting are a common side effect of chemotherapeutic agents, and this is triggered through central (chemotherapy trigger zone in the medulla) and peripheral mechanisms (serotonin receptor activation). Despite adequate treatment with antiemetic agents, 40% of oncology patients still report ongoing symptoms.
    Supplementation of cannabis has been shown to improve symptoms.
  - Analgesia [20, 21].
    Pain in cancer patients can be due to a multitude of reasons, including tumor related size/compression, chemotherapy, and the effects of radiation. Opioids have been the mainstay of treatment, but have their side effects.
    Cannabis used as an adjunct treatment for pain has been shown to be beneficial in patients with cancer.
  - Poor appetite [22]
    Tumors can release molecules that mimic natural hormones involved in satiety.
    With the approval of Dronabinol for AIDS patients, similar application for simulating appetite has been tried in cancer patients with good outcomes.
  - Tumor suppression [23–25].
    Given the antineoplastic effects of some cannabinoids, there is a growing interest in their use for treatment of the cancer itself.

Summary

Cannabis has been used for various medical and non-medical purposes. As we get to understand the effects of the different compounds in the cannabis plant, it will lead to a better understanding of the utility in medical treatment. Developing and conducting effective trials remains a challenge, but with some promising results in areas such as pain and seizure treatment, cannabis research may gain the required traction.

References

1. Fischer B, et al. Lower-risk cannabis use guidelines (LRCUG) for reducing health harms from non-medical cannabis use: a comprehensive evidence and recommendations update. Int J Drug Policy. 2021;99:103381.
2. Arnold JC. A primer on medicinal cannabis safety and potential adverse effects. Aust J Gen Pract. 2021;50(6):345–50.

3. Hussain T, et al. Cannabis sativa research trends, challenges, and new-age perspectives. iScience. 2021;24(12):103391.
4. Cooper ZD, et al. Challenges for clinical cannabis and cannabinoid research in the United States. J Natl Cancer Inst Monogr. 2021;2021(58):114–22.
5. Pisanti S, Bifulco M. Medical cannabis: a plurimillennial history of an evergreen. J Cell Physiol. 2019;234(6):8342–51.
6. Abyadeh M, et al. A proteomic view of cellular and molecular effects of cannabis. Biomolecules. 2021;11(10):1411.
7. Kanabus J, et al. Cannabinoids-characteristics and potential for use in food production. Molecules. 2021;26(21):6723.
8. ElSohly MA, et al. A comprehensive review of cannabis potency in the United States in the last decade. Biol Psychiatry Cogn Neurosci Neuroimaging. 2021;6(6):603–6.
9. Cohen K, Weinstein AM. Synthetic and non-synthetic cannabinoid drugs and their adverse effects—a review from public health prospective. Front Public Health. 2018;6:162.
10. Kosiba JD, Maisto SA, Ditre JW. Patient-reported use of medical cannabis for pain, anxiety, and depression symptoms: systematic review and meta-analysis. Soc Sci Med. 2019;233:181–92.
11. Stella B, et al. Cannabinoid formulations and delivery systems: current and future options to treat pain. Drugs. 2021;81(13):1513–57.
12. Abrams DI. The therapeutic effects of cannabis and cannabinoids: an update from the National Academies of Sciences, Engineering and Medicine report. Eur J Intern Med. 2018;49:7–11.
13. Lowe H, et al. The endocannabinoid system: a potential target for the treatment of various diseases. Int J Mol Sci. 2021;22(17):9472.
14. Jugl S, et al. A mapping literature review of medical cannabis clinical outcomes and quality of evidence in approved conditions in the USA from 2016 to 2019. Med Cannabis Cannabinoids. 2021;4(1):21–42.
15. Moreno-Rius J. The cerebellum, THC, and cannabis addiction: findings from animal and human studies. Cerebellum. 2019;18(3):593–604.
16. Bonomo Y, et al. Clinical issues in cannabis use. Br J Clin Pharmacol. 2018;84(11):2495–8.
17. Vacaflor BE, et al. Mental health and cognition in older cannabis users: a review. Can Geriatr J. 2020;23(3):242–9.
18. Minerbi A, Häuser W, Fitzcharles MA. Medical cannabis for older patients. Drugs Aging. 2019;36(1):39–51.
19. Smith LA, et al. Cannabinoids for nausea and vomiting in adults with cancer receiving chemotherapy. Cochrane Database Syst Rev. 2015;2015(11):CD009464.
20. Wang L, et al. Medical cannabis or cannabinoids for chronic non-cancer and cancer related pain: a systematic review and meta-analysis of randomised clinical trials. BMJ. 2021;374:n1034.
21. Boland EG, et al. Cannabinoids for adult cancer-related pain: systematic review and meta-analysis. BMJ Support Palliat Care. 2020;10(1):14–24.
22. Wang J, et al. Medical cannabinoids for cancer cachexia: a systematic review and meta-analysis. Biomed Res Int. 2019;2019:2864384.
23. Marcu JP, et al. Cannabidiol enhances the inhibitory effects of delta9-tetrahydrocannabinol on human glioblastoma cell proliferation and survival. Mol Cancer Ther. 2010;9(1):180–9.
24. Milian L, et al. Cannabinoid receptor expression in non-small cell lung cancer. Effectiveness of tetrahydrocannabinol and cannabidiol inhibiting cell proliferation and epithelial-mesenchymal transition in vitro. PLoS One. 2020;15(2):e0228909.
25. Ramer R, Wittig F, Hinz B. The endocannabinoid system as a pharmacological target for new cancer therapies. Cancers (Basel). 2021;13(22):5701.

Legal Aspects of Cannabis

2

Eric L. Foster and Stuart W. Ruffolo

Introduction

According to the Government of Canada, there were close to 340,000 individuals (average over 2020) who accessed cannabis for medical purposes under the *Cannabis Ac*t, which is a tenfold increase compared to 2015 [1]. The current average daily amount authorized by health care practitioners is 2 g/day [1]. In 2020, there were an average of 2330 health care practitioners associated with active registrants using cannabis, the majority of whom were in the provinces of Ontario (37%) and British Columbia (23%) [1]. These statistics highlight the speed and magnitude at which the cannabis industry is growing. As social norms around cannabis begin to change and the regulation surrounding the study of cannabis becomes less prohibitive, the use of cannabis for a variety of medical conditions is expected to grow. It is therefore important to understand the legal regime that governs medical cannabis in Canada and consider the regulations in other jurisdictions as a comparison. This chapter offers a summary of certain provisions of Canada's medical cannabis regulatory regime that are relevant to health care practitioners, hospitals, pharmacists and academics. For further information or specific regulations, the reader is encouraged to seek independent legal advice from counsel with expertise in cannabis law.

As this chapter cites various legal documents, the reference list at the end provides all the relevant Acts and Regulations.

E. L. Foster (✉)
Dentons Canada LLP, Calgary, AB, Canada
e-mail: eric.foster@dentons.com

S. W. Ruffolo
Stikeman Elliot LLP, Toronto, ON, Canada
e-mail: stuart.ruffolo@dentons.com

© Springer Nature Switzerland AG 2022
R. Valani (ed.), *Cannabis Use in Medicine*,
https://doi.org/10.1007/978-3-031-12722-9_2

The History of Medical Marijuana in Canada (See Fig. 2.1)

- Prior to the introduction of the *Cannabis Act* in 2018, the *Controlled Drugs and Substances Act* (CDSA) was the relevant legislation that prohibited the possession, production and distribution of cannabis, its active compounds and its derivatives.
  - Cannabis was included in the CDSA as is a Schedule II drug and, unless otherwise regulated, was subject to offences under the CDSA.
- In 2001, the *Marihuana Medical Access Regulations* (MMAR) were enacted under the CDSA to enable individuals who had a declaration from their medical practitioner to access dried cannabis for the treatment of their medical condition.
  - The MMAR allowed license holders to obtain dried cannabis in one of three ways:
    - Through a Personal-Use Production License (PUPL), which permitted the license holders to grow a certain quantity in their own home;
    - Through a Designated Person Production License (DPPL), which permitted a designated person to produce dried cannabis for up to two license holders; and
    - Through purchasing dried cannabis directly from Health Canada.
- In 2013, the *Marihuana for Medical Purposes Regulations* (MMPR) replaced the MMAR. However, one aspect of the regime that remained unchanged was that patients were limited to using dried cannabis.
  - In 2015, the Supreme Court of Canada held that limiting individuals with a legitimate, legally recognized medical need for cannabis to dried cannabis was unconstitutional. (R v Smith, 2015 SCC 34)
    - The Court found that the provisions of the MMPR that prohibited patients from possessing or using non-dried medicinal cannabis forced those patients to accept the risk of harm to health that may arise from the smoking of cannabis.
    - The Court reasoned that since the objective of the prohibition was the protection of health and safety, such a prohibition contradicted its objective and was therefore arbitrary. In response, the Minister of Health issued CDSA exemptions allowing for oils and fresh cannabis buds to be produced and sold by licensed producers and used and possessed by patients.
  - The MMPR system also completely reformed the regime through which individuals accessed medical cannabis.
    - The system replaced PUPLs and DPPLs with a system of government licensed producers, and mandated that dried cannabis be produced by such licensed producers.
  - The MMPRs purported to provide access to quality-controlled dried cannabis for medical purposes, produced in secure and sanitary conditions, while also providing patients with more choice of cannabis strains and licensed, commercial suppliers.
    - However, MMPR also limited patients to a single government approved contractor and eliminated the patient's ability to grow their own cannabis or to engage their own designated producer.

Fig. 2.1 Timeline of cannabis regulation in Canada

- In February 2016, the Federal Court of Canada found that requiring patients to get their cannabis only from Licensed Producers was unconstitutional because it did not provide patients with "reasonable access" to their medicine. (Allard v Canada, 2016 FC 236)
- In August 2016, the *Access to Cannabis for Medical Purposes Regulations* (ACMPR) came into force in response to the Federal Court's decision.
  - The ACMPR provided individuals who have a medical need, and a medical document from their health care practitioner, with three different ways to access cannabis:
    By registering with a licensed producer;
    By registering with Health Canada to produce a limited amount for their own medical purpose; or
    By designating a designated person to produce it for them.
- Other than being produced through the regime outlined in the ACMPR, the *Narcotic Control Regulations* (NCR) under the CDSA was another avenue through which a person was able to produce, sell, provide, transport, send, deliver or otherwise deal with cannabis.
  - The NCR authorized persons to possess limited quantities of cannabis if that person required the cannabis for their business or profession and had a dealer's license, or were a pharmacist or practitioner.
- On October 17, 2018, the *Cannabis Act* came into force. The primary significance of the Cannabis Act was that it provided for the legalization and strict regulation of cannabis for *recreational* use. However, the Cannabis Act also created a consolidated regime for cannabis for medical use by replacing the ACMPR and largely replacing the provisions that dealt with cannabis in the CDSA and NCR.
  - The *Cannabis Act* provides a stand-alone framework for controlling the production, processing, distribution sale and possession of recreational and medical cannabis in Canada.
  - Initially, the only forms of cannabis that were legalized for purchase were dried cannabis, fresh cannabis and cannabis oil. Individuals were also permitted to grow up to four cannabis plants per dwelling house. This phase of cannabis legalization is often referred to as "Cannabis 1.0."
  - On October 17, 2019, the regulations under the *Cannabis Act* were amended to allow for three new classifications of cannabis: cannabis edibles, cannabis extracts and cannabis topicals. This phase of cannabis legalization is often referred to as "Cannabis 2.0."

The Legal Framework of Medical Cannabis in Canada Today

- The *Cannabis Act* and the *Cannabis Regulations* govern the issuance of licenses and permits that authorize the importation, exportation, production, testing, packaging, labelling, sending, delivery, transportation, sale and possession of cannabis.

- While many features of the recreational cannabis regime are governed at the provincial or territorial level, the medical cannabis regime is governed exclusively at the federal level by the Minister of Health, through Health Canada.

Cannabis Products

- The following are the primary types of cannabis products that are authorized by the *Cannabis Act*:
  - **Dried Cannabis**: any part of a cannabis plant that has undergone a drying process. This does not include seeds.
  - **Fresh Cannabis**: freshly harvested cannabis buds and leaves.
  - **Cannabis Extracts**: Any substance produced through extraction from any part of the cannabis plant, or synthesizing a product that is similar to a compound in the cannabis plant. This does not include topical or edible cannabis.
  - **Cannabis Topical**: Any substance from the cannabis plant or a similar synthetic that is intended for topical use.
  - **Edible Cannabis**: Any substance from the cannabis plant or similar synthetic that is intended to be consumed. It does not include dried cannabis, fresh cannabis, cannabis plants or cannabis plant seeds.
  - **Cannabis Plant**: Any plant that belongs to the genus *Cannabis*.
  - **Cannabis Plant Seeds**: Seeds that belong to the genus *Cannabis* that can be used to grow a cannabis plant.
- Industrial hemp is defined as a cannabis plant—or any part of that plant—in which the concentration of THC is 0.3% or less in the flowering heads and leaves. (See S.1 of the *Industrial Hemp Regulations*).
  - Activities in relation to industrial hemp are largely governed by the *Industrial Hemp Regulations* under the *Cannabis Act*, which is beyond the scope of this book.

Licenses Under the Cannabis Act

- The following are the classes of licenses that authorize activities in relation to cannabis. Where applicable, the subclasses of a particular license are indicated under the applicable license class.
  - A license for cultivation (Refer to S.11 of the *Cannabis regulations*). This can be for:
      micro-cultivation;
      standard cultivation; and
      a nursery (Refer to S.14 of the *Cannabis regulations*).
  - a license for processing (Refer to S.17 of the *Cannabis regulations*);
  - a license for analytical testing (Refer to S.22 of the *Cannabis regulations*);
  - a license for medical sale (Refer to S.27 of the *Cannabis regulations*);
  - a license for research (Refer to S.28 of the *Cannabis regulations*); and
  - a cannabis drug license.

Access to Cannabis for Medical Purposes

- Part 14 of the *Cannabis Regulations* establishes a separate regime applicable to the access to cannabis for medical purposes.
- The possession limit with respect to medical cannabis is greater than that for recreational cannabis.
  - According to S.266 of the *Cannabis Regulations*, a person authorized to use cannabis for a medical condition may possess up to 30 times their daily allowance of dried cannabis or 150 g of dried cannabis, whichever is less.
  - Equivalency of dried cannabis is determined by reference to Schedule 3 of the *Cannabis Act*. See Table 2.1.
- There are four main avenues through which an individual may access medical cannabis:
  (a) By registering with licensed seller and having a medical document;
  (b) By registering to produce cannabis for personal medical purposes, or for whom cannabis may be produced by a designated person;
  (c) By obtaining cannabis as an inpatient or outpatient of a hospital; or
  (d) By being responsible for a person who is authorized to possess medical cannabis in a manner identified in items (a)–(c).

The Medical Document

- A medical document by a Health Care Practitioner to support the use of cannabis for medical purposes is required for:
  - Individuals who wish to access cannabis by registering with a licensed seller; and
  - Individuals who are registered as an authorized producer.
- In order for the Medical Document to be valid, it must contain details about the Health Care Practitioner, the patient, the relevant address, the daily quantity of dried cannabis authorized for use (in grams) and the duration of use (for a maximum of 1 year).

The Written Order

- A written order, completed by a Health Care Practitioner, states the amount of cannabis to be dispensed for the patient.

Table 2.1 Equivalency of dried cannabis to other classes

Class of cannabis for comparison	Equivalency to 1 g of dried cannabis (g)
Fresh cannabis	5
Solids containing cannabis	15
Non-solids containing cannabis	70
Cannabis concentrate	0.25

- The Written Order must be signed and dated by the Health Care Practitioner, and includes information on the prescriber, the patient, the daily quantity of cannabis to be used (in grams).
- Written Orders may also be used in connection with obtaining drugs containing cannabis.
- Written Orders may not be used to obtain cannabis from a holder of a license for sale or for registering as a registered person authorized to produce cannabis for their own medical purposes.

Holders of a License for Sale

- Once a patient has a Medical Document, they can register with licensed seller.
- Licensed sellers can register individuals as their clients upon receipt by that individual of a registration application and their Medical Document.
  - Licensed holders are responsible for verifying the validity of the Medical Document.
- Holders of a license for sale are required to transfer a Medical Document upon request or consent of the patient.
- Once a patient is registered with a licensed seller, the patient can place orders directly to the license holder.

Registration with Minister

- Another path to access cannabis for medical purposes is to register with the Minister (Registered Person).
- A Registered Person may:
  - Produce cannabis for their own medical purposes.
      They must be an adult over the age of 18 to be a Registered Person. Those under the age of 18 may only access cannabis through a Designated Person if they do not register with a holder of a license for sale.
      A Registered Person producing cannabis for their own medical purposes is permitted to cultivate, propagate and harvest cannabis as specified in their registration certificate.
      Registered Persons who are authorized to produce cannabis must take reasonable steps to ensure the security of the cannabis in their possession and the registration certificate.
  - Designate a person to produce cannabis the Registered Person.
      The regulations are similar for a designated person as with a registered person. However, Designated Persons are also permitted to send, deliver, transport or sell cannabis to their respective Registered Persons. In addition, the Designated Person can be so for up to two Registered Persons.

Hospital Patients

- An individual who is in charge of a hospital may permit cannabis products from a licensed seller to be given to a patient that has the required Medical Document or Written Order.
- If a hospital permits cannabis products to be distributed or sold, the hospital must ensure that the quantity sold to a patient does not exceed the prescribed limits. Furthermore, there are rules regarding the packaging and labeling of these items.
    - The patient must be given the current version of the *Consumer Information— Cannabis* document, published by the Government of Canada on its website [2].
- A hospital pharmacist who receives cannabis products must ensure appropriate documentation (the cannabis products, the patient, and other details).

Drugs Containing Cannabis

- Part 8 of the *Cannabis Act* contains the legal framework for drugs containing cannabis.
- A cannabis drug license may authorize the possession, production, offering to produce, sale and distribution of a drug containing cannabis.
    - A cannabis drug license does not allow the license holder to produce cannabis by cultivating, propagating or harvesting it. Instead they must obtain the cannabis from a licensed seller that is authorized to do so.
    - To be eligible for a cannabis drug license, the applicant must first hold a drug establishment license issued under the Food and Drug Regulations.
- Drugs containing cannabis may be sold or distributed to limited persons, such as holders of a research license or an analytical testing license, a pharmacist, a "practitioner" or a hospital employee.
- Persons are permitted to possess a cannabis containing drug include:
    - A pharmacist or a Practitioner if they require the drug for their business or profession;
    - A Practitioner if their possession is for emergency medical purposes only; and
    - A hospital employee or Practitioner in a hospital.
- Pharmacists are authorized to sell, distribute or administer a prescription drug that contains cannabis for which a drug administration number has been assigned under the Food and Drug Act. The drug must be dispensed in accordance with a Written Order or signed prescription.
- Practitioners are authorized to administer, sell or distribute a drug containing cannabis to an individual if the person is their patient and the drug is required for their treatment.
- Hospitals may sell, distribute or administer a prescription drug containing cannabis (having a drug an administration number under the *Food and Drug Act*) only if
    - the individual in charge of the hospital authorized it.
    - it is in accordance with a prescription or Written Order.
    - the individual is under treatment at the hospital.

Research with Cannabis

- A person who wishes to conduct research with cannabis can do so with a license issued under the *Cannabis Regulations.*
- As part of the research, the holder of the license can:
  - Alter the physical and chemical properties of the compound.
  - Administer cannabis to a research subject.
- Cannabis research activities must be described in detail in the application, along with all the research activities.
  - Detailed information regarding the type(s) of research (e.g. *in vitro, in vivo,* clinical trial, plant genetics, cannabis product development, non-cannabis product development, other), the research protocol, the quantity of cannabis required and the duration of the study will be required.
  - Other authorizations or applications may be required as well such as Experimental Studies Certificate and/or a clinical trial application filed with Health Canada. To begin the study, the researchers will need a No-Objection Letter from Health Canada.

Importation or Exportation of Cannabis for Medical or Scientific Purposes

- The *Cannabis Act* prohibits the import to and export of cannabis from Canada unless specifically authorized.
- Permits may be obtained under the *Cannabis Regulations* and are required in respect of each shipment of cannabis that is imported or exported.
- Cannabis may only imported to and exported from Canada for medical or scientific purposes.
- Only those who already hold a license issued by Health Canada are eligible for an import or export permit. Permits are granted for:
  - Importing starting materials (e.g., seeds, plants) for a new license holder;
  - Exporting to another country that has a legal regime for access to cannabis for medical purposes; or
  - Sending or receiving small quantities for scientific purposes.

Summary

There have been significant changes in the laws and regulations related to the production, sale, and use of cannabis for medical purposes in Canada. Other countries such as Australia, Denmark, several States in the US, and the UK have legalized cannabis use for medical purposes, but each with their own restrictions. As we get to understand more about the health benefits and risks of cannabis, the rules and regulations may change.

Acts and Regulations Referenced in This Chapter

Allard v. Canada, 2016 FC 236, [2016] 3 FCR 303.
Access to Cannabis for Medical Purposes Regulations, SOR/2016-230.
Cannabis Act, SC 2018, c 16.
Cannabis Regulation Act, CQLR c C-5.3.
Controlled Drugs and Substances Act, SC 1996, c 19.
Industrial Hemp Regulations, SOR/2018-145.
Marihuana Medical Access Regulations, SOR/2001-227.
Marihuana for Medical Purposes Regulations, SOR/2013-119.
Narcotic Control Regulations, CRC, c 1041.
R. v. Smith, 2015 SCC 34, [2015] 2 S.C.R. 602.

References

1. Government of Canada. Data on cannabis for medical purposes. 2021. https://www.canada.ca/en/health-canada/services/drugs-medication/cannabis/research-data/medical-purpose.html.
2. Government of Canada. Consumer information—Cannabis. 2019. https://www.canada.ca/en/health-canada/services/drugs-medication/cannabis/laws-regulations/regulations-support-cannabis-act/consumer-information.html.

Pharmacology of Cannabis

3

Rahim Valani

Introduction

Cannabis, more commonly known as marijuana, is one of the most widely used recreational substances. The United Nations Office on Drugs and Crime estimate that 4% of the adult population have used cannabis in their lifetime [1]. There are historical records dating back centuries on the medical benefits of cannabis, and with legalization of cannabis in various regions, we need to better understand the medical and recreational effects of this substance. With better knowledge of the pharmacology of the different constituent compounds in the cannabis plant, new therapeutic opportunities can being evaluated.

While medical uses of cannabis are being studied and evaluated, there is also an increase in the use of recreational cannabis and synthetic cannabinoids. Illegal and home-grown products have increased rates of contaminants such as microbes, heavy metals, pesticides, and butane all of which make consumption toxic [2, 3]. While naïve users use synthetic cannabinoids for various reasons, the effects of these synthetic compounds and contaminants are not predictable and pose a major health risk.

The Cannabis Plants

- There are three main types of *Cannabis* plant [1, 4]:
  - *Cannabis Sativa*—this is the most common type of cannabis and is grown throughout the world. It is also the most popular type in North America.
  - *Cannabis Indica* (skunk weed)—this cannabis plant is short and is known to have high concentrations of THC.

R. Valani (✉)
Department of Medicine, University of Toronto, Toronto, ON, Canada
e-mail: r.valani@utoronto.ca

© Springer Nature Switzerland AG 2022
R. Valani (ed.), *Cannabis Use in Medicine*,
https://doi.org/10.1007/978-3-031-12722-9_3

- *Cannabis Ruderalis*—this plant is primarily found in Central Asia and in the wild.
- Marijuana is technically the dried flower of the *Cannabis Sativa* plant.
  - Marijuana is also known by several names among recreational and illegal users.
    The most common alternative names are weed, herb, pot, grass, bud, ganja, Mary Jane.
  - To create a more potent cannabis product, small scale and home users use butane or other solvents to develop concentrated resins of the compounds. Unfortunately, these are not always effective, and contaminants are common.
  - The consistency of the extract is how the product is recognized (shatter, honeycomb, crumble wax, budder, and earwax) [5].
- The cannabis plant contains over 100 identified cannabinoids, and other compounds including hydrocarbons, terpenes, flavonoids, and non-cannabinoid phenols. The effect of each of these compounds is not fully understood and is the focus of active research.
  - In addition, different parts of the cannabis plant contain different concentration of the cannabinoids and other substances. See Table 3.1 [6].
- Phytocannabinoids are natural molecules from plants with an affinity for the cannabinoid system. The two most common cannabinoid compounds are Δ9-Tetrahydrocannabinol (THC) and Cannabidiol (CBD) [2].
  - THC concentrations of cannabinoids from confiscated samples have increased in the past two decades, from 3.8% in the 1990s and 12.2% in 2014.
  - In the case of extracts, the THC concentration have been noted to be as high as 80%.
  - While THC and CBD are the primary focus in the medical community, other classes of cannabinoids include [6]:
    Δ9-Tetrahydrocannabindiol.
      - Neuroprotective effects, anticonvulsant, and muscle relaxant.
    Δ9-Tetrahydrocannabinol acid.
      - Anticonvulsant effects, and PPARγ agonist.
    Δ8-Tetrahydrocannabinol.
    Cannabigerol.
      - Effects on pain modulation, inflammation, and heat sensitization.

Table 3.1 Concentration of cannabinoid for different parts of the cannabis plant [1, 6]

Cannabis plant	Cannabinoid %	Other
Fan leaves	0.05	Terpinoids and flavonoids
Stem	0.02	Cellulose
Roots	0	Terpinoids
Unfertilized flower	Up to 30	Terpinoids (up to 4%)
Fertilized flower	Up to 13%	
Seeds	0	Essential fatty acids (35%)
Capitate glandular trichomes	Up to 60%	Terpinoids (up to 8%)

  Cannabichromene.
  Cannabinodiol.
  Cannabielsoin.
  Cannabicyclol.
  Cannabinol.
  Cannabitriol.
- The THC/CBD ratio defines the potency and psychoactive effects of the cannabis product. (See section on pharmacokinetics which outlines the different effects these compounds have.)
- Various solvents have been used to extract the different chemical compounds from the cannabis plant. These include [2]:
  - Chemical solvents (petroleum ether, ethanol, naptha)—these can leave unwanted residues and have their own toxic effects.
  - Liquid carbon dioxide (CO_2).
  - Organic products (olive oil, coconut oil). These tend to be labelled as "organic" or natural.
- At the same time, there are several contaminants in addition to the solvents that are harmful. Some of these include [3]:
  - Microbes—usually bacteria and fungi.
    Most of this occurs from improper preparation and storage of cannabis and cannabis products.
  - Heavy metals.
    Heavy metal toxicity can occur from one of three ways:
    - Bioaccumulation based on where the plant is grown (fertilizers in the soil having metals such as cadmium).
    - Cross-contamination during processing.
    - Post-processing adulteration—heavy metals added so as to add wight to the compound and increase its value.
  - Pesticides.
    Pesticide consumption has been shown to result in malignancy and developmental issues. They have also been shown to affect the reproductive organs, neurologic system, and endocrine system.

The Endocannabinoid System (See Fig. 3.1) [6–11]

- The endocannabinoid system in humans is activated by endogenous bioactive lipids that bind to the cannabinoid receptors. The activity of the cannabinoid is rapidly terminated either through cellular uptake or via intracellular degradation.
- There are two main native endocannabinoids compounds:
  - *N*-arachidonoylethanolamine (Anandamide)
    Anandamide is seen primarily in brain tissue and is more selective to CB_1 receptors.

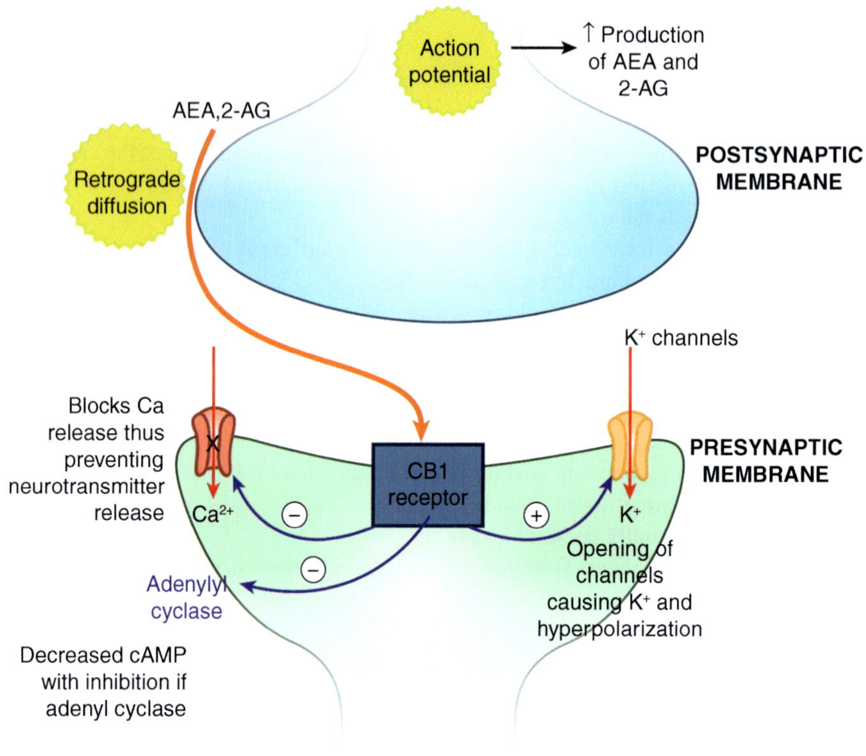

Fig. 3.1 The endocannabinoid system

The inhibitory constants, K_i, defines the binding affinity for a substance to the receptor. The K_i for Anandamide for the cannabinoid receptors are:

- CB_1 K_i = 89 nM
- CB_2 K_i = 371 nM

Anandamide is primarily metabolized by Free Fatty Acid Hydroxylase (FFAH). A secondary pathway through COX-2 converts it into proalgesic prostamides:

$$ANADAMIDE \xrightarrow{FFAH} ETHANOLAMINE + ARACHIDONIC\ ACID$$

- 2-Arachidonoylglycerol (2-AG)

    2-AG is found mainly in the GI tract, and is metabolized by Monoacylglycerol Lipase:

$$2-AG \quad \xrightarrow{Monoacylglycerol\ Lipase} \quad GLYCEROL + ARACHIDONIC\ ACID$$

    The inhibitory constants for 2-AG are:
    - $CB_1\ K_i = 472$ nM
    - $CB_2\ K_i = 1400$ nM

Endocannabinoid Receptors

- There are two main endogenous endocannabinoid receptors in the body, namely CB_1 and CB_2. These receptors are 7-transmembrane G-protein coupled receptors.
    - CB_1 receptors:

        The CB_1 receptor is highly expressed in neuronal tissue, both at the pre- and post- synaptic neurons.

        Activating the CB_1 receptor inhibits adenylate cyclase through inhibition of N-, Q-, and L-type calcium channels. This results in decreased neurotransmitter release from the synapses.

        The CB_1 receptor has been found in the following tissue/organs:
        - Nucleus of solitary tract, which is responsible for antiemetic effects.
        - Hypothalamus.
        - Motor cortex.
        - Basal ganglia.
        - Cerebellum.
        - Motor neurons in the spinal cord.
        - Eyes.
        - Sympathetic ganglia.
        - Enteric nervous system.
        - Immune system—bone marrow, thymus, spleen, tonsils.
    - CB_2 receptors:

        These receptors are located mainly in immune system—bone marrow, thymus, spleen, tonsils, T lymphocytes, B lymphocytes, Natural Killer cells, PMNs.

        Activation of this receptor inhibits adenylate cyclase activation and stimulates MAP kinases.

        Studies have shown upregulation of CB_2 receptors during inflammation, which helps explain some of the anti-inflammatory properties seen with CB_2 receptor agonists.
- Other receptors: Theses are classified as orphan receptors because they have shown some activity in the endocannabinoid system. These include [12–14]:
    - Transient receptor potential of vanilloid type 1 and 2 (TRPV1 and TRPV2).
    - Transient receptor potential of ankyrin type 1 (TRPA1).

- GPR 55 receptor. This is sometimes referred to as the CB_3 receptor.
- GPR18 receptor.
- GPR19 receptor.
- GPR55 receptor.
- GPR119 receptor.

Pharmacokinetics of Phytocannabinoids [8, 9, 15]

- Δ9-Tetrahydrocannabinol (THC) and Cannabidiol (CBD) are both lipophilic compounds.
- They undergo first pass metabolism when ingested, thus reducing their bioavailability (5–20% for THC and 6–19% for CBD).
- Adsorption, metabolism, distribution, and excretion varies by age, other medications, and comorbidities. Therefore, a start low and go-slow approach is recommended (see Chap. 4).
- Exogenous cannabinoids are metabolized by the cytochrome P450 enzymes. The most common enzymes area CYP1A2, CYP2C9, CYP2C19, and CYP3A4.
- While cannabis use is generally well tolerated, the effects of co-administering with other medications should be thoroughly reviewed given the potential metabolic and pharmacokinetic effects. Specific consideration should include:
  - Metabolic enzyme interaction.
    Rifampicin is a CYP3A4 inducer and can reduce peak plasma concentrations of CBD.
    CBD is a potent inhibitor of CYP2C19. This enzyme is also responsible for the metabolism of clobazam. Therefore, the patient must be carefully monitored and the dose of clobazam will need to be modified.
  - Pharmacokinetic interactions.
    Other compounds or medications may interfere with absorption, distribution, metabolism, and excretion of cannabis or the other compound.
  - CBD inhibits p-glycoprotein mediated drug transport.

Δ9-Tetrahydrocannabinol (THC)

- This is the most studied compound from the cannabis plant.
- THC binds primarily to CB_1 receptors.
  - THC inhibits T-type Calcium channels, Potassium (Kv1.2) channels, Sodium channels, and conductance between cells.
- Since there are other compounds (such as synthetic cannabinoids) that have greater effects on the CB receptors, THC is considered a partial agonist for the endocannabinoid receptors (see Table 3.2 on pharmacokinetic parameters for THC and CBD).
- It is highly protein bound, with a half life of 30 h, and less than 1% is eliminated in its native form.

Table 3.2 Pharmacokinetics of THC and CBD [6]

Pharmacokinetics	THC	CBD
Volume of distribution (L/kg)	7.5–8.9	32
CB_1 K_i (nmol/L)	5.05–80.3	
CB_2 K_i (nmol/L)	1.73–75.3	
Half life (h)	30	9–32

- Metabolism follows non-linear kinetics, making it difficult to determine time of ingestion based on serum or urine levels.
- THC is primarily metabolized by isoenzymes CYP2C9, CYP2C19, and CYP3A4.
  - THC can also affect certain CYP450 enzymes:
    It is known to inhibit CYP- 2C9, 3A4, and 2D6 and induces CYP1A2.
- THC is hydroxylated to 11-hydroxy-Δ9-Tetrahydrocannabinol (Δ9-THC-OH) by cytochrome p450. It is Δ9-THC-OH that is the most active metabolite.

$$THC \xrightarrow{P450\,system} \Delta 9 - THC - OH + 11 \rightarrow 11 - nor - 9 - carboxy$$
$$-THC\left(biologically\ inactive\right) \xrightarrow{Glucoronidation} THC - COOH - glucoronide$$

- 65% is excreted in feces and 13% in urine.
  - The primary urinary cannabinoid is THC-COOH-glucuronide.
- Post consumption, 80–90% of cannabinoids are usually excreted within 5 days.
- THC is also converted to Cannabinol through non-enzymatic oxidation. This by-product is most commonly seen with prolonged storage of the compound [6].
  - Cannabinol has only 25% the potency of THC.
- Monitoring of THC use: (see Chap. 5)
  - THC can be detected for approximately 6 h in the blood after smoking. Therefore, the presence of THC in a blood sample suggests recent ingestion.
  - THC-COOH blood concentration ≤3 µg/L is considered a marker of occasional intake, while ≥40 µg/L is a marker of nearly daily use.
  - There are no good urine markers for recent cannabis intake as THC and 11-OH-THC can be detected up to 24 days post use.
- There is a dose-dependent response resulting in different effects:
  - <1 µM
    Activates various glycoproteins and peroxisome proliferation.
    Enhances release of calcitonin gene-related peptide.
    Potentiates glycine-ligated ion channels.
    Antagonizes serotonin (5-HT_{3A}) ligand.
  - 1–10 µM
    Activates TRPV and TRPV4 channels.
    Potentiates β-adrenoreceptors.
    Displaces opiates from µ-opioid receptor.

- The physiologic effects of THC include:
  - Pain modulation—helps with chronic pain modulation.
  - Decreased spasticity—effects seen in patients with Multiple Sclerosis.
  - Sedation.
  - Appetite changes—increased appetite and decreased nausea/vomiting when used appropriately.
  - Mood changes.
  - Antioxidant activity.
  - Antipruritic agent (seen in cholestatic jaundice).
  - Anti-inflammatory regulation.
  - Possible reduction in intra-ocular pressure.

Cannabidiol (CBD)

- CBD is the most abundant phytocannabinoid found in European hemp.
- It is an agonist at the TRPV1 and 5-HT_{1A} receptors, and binds primarily to the CB_2 endocannabinoid receptor [6].
- CBD is metabolized primarily by isoenzymes CYP2C19 and CYP3A4. Other isoenzymes that metabolize CBD include CYP1A1, CYP1A2, CYP2C9, and CYP2D6.
- CBD also inhibits certain Cytochrome P450 enzymes—CYP 2C19, 3A4, 2D6
  - Studies have shown increased concentration of Clobazam in patients with CBD, given the metabolism of Clobazam is dependent on two of these isoenzymes.
- CBD is highly protein bound, and has a variable half life between 9 and 32 h.

$$Cannabidolic\ Acid \rightarrow CBD \xrightarrow{Hydroxylation} 7-OH-CBD$$

- 7-hydroxy-CBD (7-OH-CBD) undergoes further hepatic metabolism and is eventually excreted through feces (less through the urine).
- CBD counteracts some of the untoward effects of other cannabinoids such as anxiety, tachycardia, hunger, and sedation.
- The physiological effects of CBD include:
  - Anti-inflammatory.
  - Antioxidant.
  - Antipsychotic effects.
  - Immunosuppression.

Terpenoids [6]

- Terpenoids are compounds found in the cannabis plant and help with predator protection and attract pollinators. They are stored in the hair-like protrusions (glandular trichomes) [2].

- They are produced in the glandular trichomes, and reflect the environmental conditions of the plant (growing conditions, surrounding environment).
- Most terpenes show low toxicity, with an LD_{50} dose ≥ 5000 mg/kg.
- The pharmacological effects of terpinoids are not fully understood. However, they are seen in various preparations depending on the extraction methods and mode of intake of the cannabis product.
- β-Myrcene is the most prevalent terpinoid and is responsible for the sedation seen in most commercial preparations.
  - It has a musky fragrance to it.
  - It may possess some antioxidant and anti-carcinogenic properties.
- α- and β-pinene
  - They have a more pine fragrance.
  - It is postulated to have antiseptic effects.
- Other terpinoids include:
  - D-Limonene—this is the parent compound for the monoterpinoids of the cannabis plant.
  - β-Ocimene—most common monoterpene.
  - α-Terpinene and γ-Terpinene.
  - α-Terpineol.
  - α-Pinene and β-Pinene.
  - Linalool.
  - Camphene.
  - Terpinolene.

Synthetic Cannabinoids

- Synthetic cannabinoids are manufactured cannabinoid products. There were more than 160 reported synthetic cannabinoids reported between 2008 and 2016, and the numbers keep increasing [16].
  - Synthetic cannabinoids include HU-210, HU211, JWH-018, AB-FUBINACA, 5F-PB-22, AB-PINICA, BB-22, EG-018, to name a few.
- Most of the synthetic cannabinoids are developed for recreational and illegal use. They go by various street names:
  - K-2, Black mamba, crazy clown, spice, Spice2, Smoke, and Summit.
- Many of the synthetic cannabinoids are developed with the aim of increasing the euphoric effects of cannabis. As a result, their effects on the endocannabinoid receptors are much higher than THC.
  - JWH-018 has 4 times the affinity for the CB_1 receptors compared to THC, and 20 times the affinity for CB_2 receptors.
- The lack of detection on a routine urine toxicology screen makes it more attractive for users who do not wish to disclose use for employment or legal screening.
- Side effects and potency can vary depending on the chemical composition, type of extraction, and adulterants (see Table 3.3).

Table 3.3 Side effects commonly seen with synthetic cannabinoids

Systems	Effect
Neuropsychiatric	Psychosis, agitation, aggression, hallucinations, panic attacks, suicidal ideation
Cognitive	Memory alteration, amnesia, attention difficulties
Cardiovascular	Tachycardia, hypertension, arrhythmias
Neurologic	Dizziness, seizure, sensory alteration, hypertonicity
Gastrointestinal	Nausea, appetite changes

- There are two synthetic cannabinoids in the global market approved for medical use:
  - Nabilone (Cesamet, Canemes)

    Provided as capsules that contain the synthetic cannabinoid (analogue of THC).

    The product was approved by the FDA in 1985 for the main indication of chemotherapy associated nausea/vomiting.
  - Dronabinol (Marinol, Syndros)

    This product is available as an oral capsule/solution.

    It was approved by the FDA in 1985 for the following indications:
    - Anorexia associated with weight loss in patients with AIDS.
    - Chemotherapy associated nausea/vomiting.
- There are also two cannabis-derived preparations (from the plant) approved for medical use:
  - Nabiximols (Sativex)

    The product is an oromucosal spray that contains both THC and CBD in a 1:1 ratio (2.5 mg of CBD and 2.7 mg of THC).

    It is being used for spasticity in patients with multiple sclerosis.
  - Epidolex

    Approved by the FDA for the treatment of two specific types of epilepsy: Lennox-Gastaut syndrome and Dravet syndrome.

Routes of Administration (See Table 3.4) [8–10, 17]

- Canabinoids can be administered in several ways. However, the bioavailability varies depending on several factors, including the preparation of the product, the route of administration, metabolism/excretion, and other user-dependent factors (age, gender, comorbidities, other medications).

Inhalation [3]

- Inhalation delivers cannabinoids to the alveolar capillaries efficiently. This route avoids any first pass metabolism, which permits other toxins to bypass the liver and reach systemic circulation.

Table 3.4 Pharmacokinetics based on routes of administration [10]

Route	Onset (min)	Duration (h)	Bioavailability	Comments
Smoking/vaping	5–10	2–4	2–56%	Rapid onset Great for episodic symptom relief By-products of combustion present
Oral	60–180	6–8	Up to 20%	Convenient formulation
Topical	Variable	Variable		Limited use for local symptoms Issues related to absorption

- The product can be rolled into cigarettes (joints), in pipes, water pipes (bongs), or cigar wraps (blunts).
- Routes of inhalation:
  - Smoking.
    Smoking is the most common route of administration.
    The product is heated to 600–900 °C and inhaled.
    - Unfortunately, combustion also produces tar, polycyclic hydrocarbons, and carbon monoxide, which have serious health effects.
  - Vaporization—heating but not to the point of burning the compound.
    Vaporization requires the product to be heated between 160 and 230 °C. While this method has decreased carbon monoxide levels, it still contains the other by-products of combustion.
  - Dabbing—vaporization of concentrated butane hash oil with a blowtorch [5].
- The carcinogenic load of smoked non-medical cannabis is high.
  - Heavy metals such as cadmium and arsenic and some pesticides are highly volatile and convert into carcinogenic by-products during pyrolysis.
- Dosing with this route is highly variable and depends on a number of factors:
  - Time of the last meal.
  - The number of puffs taken.
  - Duration and depth of inhalation.
  - How long the breath was held for.
  - Duration of exhalation.
  - Time in between puffs.
  - Lung function of the user.
  - Temperature of the vaporizer.

Oral [3]

- This is the most convenient form of administration. The product can come in several forms such as capsules, lozenges, and oils.
- Absorption is from the highly vascularized mucosa.
- Dosing is also easier with capsules/formulated liquids given the consistency of the manufacturing process.

- Cannabis compounds can also be added to other food products such as brownies, cookies, and candies to make it more palatable.
- With oral use, cannabis is degraded by gastric acid and also subject to first pass metabolism. Both of these decreases the systemic bioavailability of the compound.
    - Bioavailability can be as low as 6%.
    - Co-ingestions with a fatty meal can increase bioavailability given the lipophilic nature of cannabis.
- Formulations can be of either THC, CBD or mixed and with varying concentrations. Examples within each category include:
    - THC synthetic formulations: Dronabinol.
    - THC derivative: Nabilone.
    - CBD: Arvisol, CardiolRx, and Epidiolex.
    - Mixed THC and CBD: Nabiximols.

Topical

- Dermal therapy can be either transdermal (for systemic effects) or topical (for localized effects).
- Dosing is more difficult with this method given variability in absorption that is either product (formulation, type of product, solubility, duration of contact) or patient dependent.
- To improve permeability across the skin, there are several options undergoing research:
    - Microneedles that create pores to increase
    - Microdermal abrasion.
    - Adding molecules that work as enhancers.
    - Transdermal patches.

Special Populations

- The effects of cannabinoids in the general population are being studied, and ongoing research continues to provide guidance on the benefits, risks, uses, and dosing.
- The use and dosing in certain populations is not well known, and caution must be taken especially for the following groups:
    - Pregnant women/breast-feeding mothers (see Chap. 15):
        THC readily crosses the placenta leading to fetal exposure.
        Cannabinoids are also expressed in breast milk.
    - Children (see Chap. 14)
        It is unclear what the developmental effects are for children on cannabis for medical reasons.
        Further studies may shed light on the uses and dosing.

- Elderly patients.

  There is increasing use of marijuana in adults >50 years old.

  Associated factors related to use include male gender, unmarried, multiple chronic diseases, psychological stressors, use of other substances (alcohol, tobacco, other illicit substance use).

  Given the vulnerability of this population, careful titration to help observe for the desired effect but minimizing adverse reactions.

Summary

The cannabis plant consists of many compounds, and we continue to learn more about the effects of these with ongoing research. Being aware of the pharmacokinetic and pharmacodynamic elements of the different products is important to help the clinician with dosing, need for monitoring with other medications, as well as side effects.

References

1. United Nations Office on Durgs and Crime. Recommended methods for the identification and analysis of cannabis and cannabis products; 2009.
2. Grof CPL. Cannabis, from plant to pill. Br J Clin Pharmacol. 2018;84(11):2463–7.
3. Dryburgh LM, et al. Cannabis contaminants: sources, distribution, human toxicity and pharmacologic effects. Br J Clin Pharmacol. 2018;84(11):2468–76.
4. McPartland JM. Cannabis systematics at the levels of family, genus, and species. Cannabis Cannabinoid Res. 2018;3(1):203–12.
5. Al-Zouabi I, et al. Butane hash oil and dabbing: insights into use, amateur production techniques, and potential harm mitigation. Subst Abus Rehabil. 2018;9:91–101.
6. Russo EB, Marcu J. Cannabis pharmacology: the usual suspects and a few promising leads. Adv Pharmacol. 2017;80:67–134.
7. Lu Y, Anderson HD. Cannabinoid signaling in health and disease. Can J Physiol Pharmacol. 2017;95(4):311–27.
8. Lucas CJ, Galettis P, Schneider J. The pharmacokinetics and the pharmacodynamics of cannabinoids. Br J Clin Pharmacol. 2018;84(11):2477–82.
9. Mouhamed Y, et al. Therapeutic potential of medicinal marijuana: an educational primer for health care professionals. Drug Healthc Patient Saf. 2018;10:45–66.
10. MacCallum CA, Russo EB. Practical considerations in medical cannabis administration and dosing. Eur J Intern Med. 2018;49:12–9.
11. Morales P, Reggio PH, Jagerovic N. An overview on medicinal chemistry of synthetic and natural derivatives of cannabidiol. Front Pharmacol. 2017;8:422.
12. Guerrero-Alba R, et al. Some prospective alternatives for treating pain: the endocannabinoid system and its putative receptors GPR18 and GPR55. Front Pharmacol. 2018;9:1496.
13. Morales P, Reggio PH. An update on non-CB(1), non-CB(2) cannabinoid related G-protein-coupled receptors. Cannabis Cannabinoid Res. 2017;2(1):265–73.
14. Sarzi-Puttini P, et al. Cannabinoids in the treatment of rheumatic diseases: pros and cons. Autoimmun Rev. 2019;18(12):102409.
15. Alsherbiny MA, Li CG. Medicinal cannabis-potential drug interactions. Medicines (Basel). 2018;6(1):3.
16. Papaseit E, et al. Cannabinoids: from pot to lab. Int J Med Sci. 2018;15(12):1286–95.
17. Brown SJ, et al. Use of cannabis during pregnancy and birth outcomes in an Aboriginal birth cohort: a cross-sectional, population-based study. BMJ Open. 2016;6(2):e010286.

Patient Assessment and Dosing Recommendations for Cannabis

4

Rahim Valani

Introduction

Cannabis has been shown to be helpful in the active management of several medical conditions. As more studies continue to broaden the scope of use, it is important for clinicians to understand how to prioritize patient care for new and ongoing uses. This chapter presents an overview of best practices related to the initial assessment of patients who wish to consider cannabis for treatment, contraindications to cannabis use, dosing, titration, and the required follow up.

- There are many strains of cannabis available with varying concentrations of chemical compounds including phytocannabinoids. The relative proportions of these components determine the effects and adverse reactions with a particular chemical variety/mixture (primarily Δ9-Tetrahydrocannabinol and Cannabidiol).
- There are many chemical compounds found in the cannabis plant, which include:
  - Monoterpenoids.
  - Phytocannabinoids.
  - Myrcene (analgesia and sedation).
  - Limonene (antidepressant and immune regulation).
  - Pinene (acetylcholine esterase inhibitor).
  - Beta-caryophllene (anti-inflammatory and analgesia).
- Producers of cannabis may use different strains and identify their products by the composition of Δ9-Tetrahydrocannabinol (THC) and Cannabidiol (CBD). It is important to realize that other constituents also have effects that contribute to effects on the patient's physiology.

R. Valani (✉)
Department of Medicine, University of Toronto, Toronto, ON, Canada
e-mail: r.valani@utoronto.ca

© Springer Nature Switzerland AG 2022
R. Valani (ed.), *Cannabis Use in Medicine*,
https://doi.org/10.1007/978-3-031-12722-9_4

- Type I cannabis—THC predominant.
- Type II cannabis—mixed THC and CBD.
- Type III cannabis—predominantly CBD.
- The entourage effect postulates that taking the combination of compounds (e.g. smoking the cannabis plant) may provide more beneficial effect as opposed to the isolate or discrete compounds like THC or CBD.
- There is varying evidence for the use of cannabis for a multitude of medical conditions. While there is evidence is supportive for the following conditions, ongoing studies are showing promise in other conditions as well [1–3]:
  - Strong evidence:
    Chronic pain. (See Chap. 11.)
    Spasticity due to multiple sclerosis. (See Chap. 9.)
    Chemotherapy induced nausea and vomiting.
    Intractable seizures with Dravet and Lennox-Gastaut syndromes. (See Chap. 9.)
  - Moderate evidence:
    Improving sleep in patients with chronic pain.
    Multiple sclerosis. (See Chap. 9.)
    Fibromyalgia. (See Chap. 10.)

Initial Clinical Assessment [4–7]

- The initial clinical assessment helps to identify if a patient is suitable candidate for medical cannabis. This is an important process to help appreciate the current management of the patient as well as understand their expectations should they start cannabis.
- An in-person interview helps establish the physician–patient relationship and whether this patient would benefit from cannabis use. Any conflicts of interest should be addressed prior to this engagement.
- During the first session, a complete history and physical examination must be completed. This helps identify any risks to the patient. The history should include:
  - Past medical history.
    Cardiovascular and respiratory history.
    Mental health history.
    Substance abuse history.
  - Any prior surgeries.
  - A current list of medications, including prescription medications, over the counter medications, as well as any herbal or natural supplements.
    Avoid cannabis in patients taking opioids or benzodiazepines. Co-ingestion may result in additional impairment and sedation which can pose a risk for safety sensitive tasks such as operating a motor vehicle or occupational tasks with heavy machinery.

- A personal or family history of mental health illness or addiction.
  In particular, ask about any history of schizophrenia.
- Prior cannabis use to determine if the patient has built a tolerance.
- A complete work history to identify any work related concerns.
  Working with heavy machinery, safety sensitive jobs, driving a motor vehicle and working with children should be carefully reviewed. Cannabis should not be offered to any patient who is in a safety sensitive occupation.
    - It is important to advise patients to avoid driving especially earlier in the course of cannabis use.
    - They can start to drive if they have reached a stable dose for 5–7 days, are not on any other medications that can impair driving, and do not consume alcohol.
    - Cannabis should not be used by professional drivers such as those operating a taxi, bus, or ambulances.
  Consider a urine toxicology screen to assess for any drugs of abuse or other substances that may interfere with cannabis use.
- The patient should be screened early for any contraindications to cannabis. The recommendations by Health Canada and other regulatory bodies are to avoid cannabis use in the following patients:
  - Any patient under the age of 18 years.
  - Patents with cardiopulmonary disease.
  - Those with respiratory issues (asthma, COPD).
  - Severe renal or liver disease.
  - Patient with a mental health history or family history of schizophrenia.
  - Mothers who are breastfeeding.
  - Patients with a history of substance abuse (alcohol, other drugs or psychoactive substances).
- All potential patients must be screened for addictions risk given the risk of cannabis abuse. The following addiction risk instruments have been validated and should be used at the initial visit and regular intervals:
  - The drug abuse screening test (DAST).
  - Opioid risk tool.
  - CAGE questionnaire for alcohol dependency.
- If the patient is a suitable candidate for medical cannabis and decides to proceed with using it, the physician should obtain informed consent to begin treatment. The consent should be documented clearly in the chart.
  - The consent should include:
    The expected benefits of the treatment based on current evidence and practice.
    Any material risks, including:
      - Precipitation of psychotic symptoms.
      - Cognitive impairment which can affect their fitness to engage in certain activities and occupations.
      - Impact on safety sensitive occupations.

- • The impact on insurance coverage.
- • Any other risks that are material and significant for the patient.
  Suitable alternatives to using cannabis must be explored.
- • This is particularly important if:
  - The patient has not tried mainstream medical therapy.
  - There is no strong evidence to the use of cannabis for a particular condition that the patient is suffering from.
- The decision to prescribe cannabis should be based on a shared decision model between the physician and the patient. The patient should adhere to the terms of a contract/agreement that prevents misuse or abuse of the product. The contract should be introduced at the first visit and must include:
  - That the patient will only present to one physician for the prescription.
  - The cannabis will only be used for the indication it has been prescribed for.
  - They will avoid any illicit substance use.
  - The patient will not share their medications with anyone, nor will they borrow from someone else.
  - The cannabis will be taken as prescribed.
  - The patient should keep a cannabis dosing diary.

Dosing Recommendations [1, 3–6]

- Dosing caveats:
  - The initial dose should be started at low levels and titrated slowly (sometimes as long as 2 weeks between increasing the dose) until the desired effect is obtained.
  - Slow titration prevents THC side effects such as tachycardia and dizziness form being overwhelming early in the treatment and allow some tolerance. With pure CBD isolates or lower concentration of THC, these side effects can be minimal.
  - Inform the patient that it can take time to see the effects of cannabis and not to hasten the titration process.
  - For chronic conditions, the optimal goal would be to have the patient on long acting oral preparations and use inhalation treatment as needed (PRN dosing).
  - Most patients will use anywhere between 1 and 3 g of cannabis a day.
    Dosing is individualized and should be titrated accordingly.
    Less than 5% of patients will need greater than 5 g/day. If a patient requires a dose greater than 5 g/day, the utility of this treatment should be questioned.
- It is important to inform the patient that most adverse events are early in the course of treatment and transient. Many of these will subside.
  - The most common side effects to look for are dizziness, somnolence, vertigo, and xerostomia.
  - If side effects persist, the patient should be advised to either decrease their dose of cannabis to a level where the side effects are tolerable or consider a different strain.

Table 4.1 THC versus CBD for specific medical conditions

Medical condition	Dosing recommendations
Chronic pain	10–40 mg of CBD
	2.5–40 mg of THC
Nausea/vomiting from chemotherapy	Up to 90 mg/day of THC
Appetite stimulation for patients with HIV/AIDS	Up to 10 mg/day of THC
Refractory seizures	1–20 mg/kg/day of CBD

- For patients who are on compounds that are primarily THC, they should use the product at nighttime to limit adverse events.
  - Once daily dosing is recommended during the escalation phase. This can be adjusted to twice daily dosing should the patient require higher doses or need symptom relief during the daytime.
- Dosing varies depending on the condition and whether a predominantly THC, CBD, or balanced formulation would be most effective.
  - Table 4.1 provides some guidance on the preferred agent of choice based on the medical condition.

Administration

- Oral dosing.
  - The oral route is preferred due to ease of ingestion. The bioavailability is poor due to first pass metabolism.
      High fat meals taken at the time of ingestion increases the bioavailability, especially of CBD given its lipophilic properties.
  - It is recommended that a first-time users wait at least 3 h before taking a second oral dose.
  - Titration of oral cannabis is much easier compared to the other routes of administration.
  - Once daily dosing:
      Day 1 and 2: 2.5 mg of THC equivalent at bedtime.
      - Increase by 1.25–2.5 mg of THC at a minimum interval of 2 days as long as the prior dose is tolerated (see caveat above for going slow with the titration).
      - Titrate to desired effect and using the minimal dose required to alleviate symptoms and minimize side effects.
      Twice daily dosing:
      - Day 1: 2.5 mg of THC equivalent at bedtime.
      - Once tolerated and a minimum of 2 days have passed, then increase to 2.5 mg twice a day.
      - Continue to increase until desired effect is achieved and using the minimal required dose.
      The mean dose for oral cannabis is 3.4 g/day.

- Vaporizing
  - It is recommended that a first-time users wait at least 15–20 min between inhalations.
  - Start with one inhalation and wait 15 min.
  - Repeat until desired effect or if the side effects occur. If the side effects are not tolerated well, then either decrease the dose or switch to another chemovar.
  - The mean dose is 3.0 g/day for inhaled cannabis.

Follow Up Appointments [5, 6, 8, 9]

- The physician who prescribes cannabis is responsible for following up on their patient.
  - Patients should be followed up at regular intervals (every 3–12 months, with shorter interval at the start of the treatment or if the patient has a higher risk of complications).
- At each visit, the following should be reviewed:
  - The patient's clinical condition and any ongoing or new symptoms.
  - Any changes to their medical history or medications.
    Review the patient medications and ensure there are no cannabis-drug interactions (see Table 4.2 for a partial list).
  - Use of any illicit substances.
  - The effects of cannabis use.
    Document the patient's symptoms and any relief so far.
    - When was the last dose change (increase or decrease)?
      - The dosing can be titrated up or down accordingly based on symptoms and side effects.
    Has the patient used the cannabis for any other reasons other than for that which it was medically prescribed?
  - Ensure the patient does not have any contraindications for continued cannabis use.
  - It is good practice to review the contract with the patient again so as to avoid any misunderstandings.

Table 4.2 Examples of metabolic interactions to consider when prescribing cannabis

Enzyme	Example drug	Effect
CYP3A4	Inducer: phenytoin	Inducer: decreased CBD availability and possibly its effectiveness
	Inhibitor: ketoconazole	Inhibitor: increased CBD availability
CYP2C19	Inducer: carbamazepine, phenytoin, phenobarbital.	Inducer: decreased CBD availability
	Inhibitor: fluoxetine	Inhibitor: increased CBD availability

Summary

Appropriate screening and dosing of cannabis is important when prescribing patients cannabis for managing medical conditions. This chapter highlights the need for thorough screening and provides dosing recommendations.

References

1. Mouhamed Y, et al. Therapeutic potential of medicinal marijuana: an educational primer for health care professionals. Drug Healthc Patient Saf. 2018;10:45–66.
2. Allan GM, et al. Simplified guideline for prescribing medical cannabinoids in primary care. Can Fam Physician. 2018;64(2):111–20.
3. MacCallum CA, Russo EB. Practical considerations in medical cannabis administration and dosing. Eur J Intern Med. 2018;49:12–9.
4. Government of Canada. Information for health care professionals. Cannabis (marihuana, marijuana) and the cannabinoids. 2018. https://www.canada.ca/content/dam/hc-sc/documents/services/drugs-medication/cannabis/information-medical-practitioners/information-health-care-professionals-cannabis-cannabinoids-eng.pdf.
5. College of physicians and surgeons of Alberta. Cannabis for medical purposes; 2014. 2021. https://cpsa.ca/wp-content/uploads/2020/06/AP_Cannabis-for-Medical-Purposes.pdf.
6. Wedman-St. Louis, B. Cannabis: a clinician's guide. Boca Raton: Taylor & Francis Group; 2018.
7. Häuser W, et al. European Pain Federation (EFIC) position paper on appropriate use of cannabis-based medicines and medical cannabis for chronic pain management. Eur J Pain. 2018;22(9):1547–64.
8. Government of Canada. Data on cannabis for medical purposes. 2021. https://www.canada.ca/en/health-canada/services/drugs-medication/cannabis/research-data/medical-purpose.html.
9. Brown JD, Winterstein AG. Potential adverse drug events and drug-drug interactions with medical and consumer cannabidiol (CBD) use. J Clin Med. 2019;8(7):989.

Ola Z. Ismail and V. Tony Chetty

Introduction

With the increased use of cannabis, there is a need to find accurate ways to measure levels of cannabinoids in the body. This may be for research, occupational, or legal reasons. There are several analytical methods used to identify the molecular constituents of a substance. The type of analytical method used depends on the physical and chemical properties of the compound as well as what particular compounds need to be identified. While advances in analytical and medical chemistry have provided more details on the different molecules in the cannabis plant, the exact mechanism of action for some of these products are still undetermined. Furthermore, clinical testing is currently limited to the more common compounds such as Δ9-Tetrahydrocannabinol (THC) and Cannabidiol (CBD).

Cannabinoid testing in the clinical laboratory usually aims to detect THC or its metabolites depending on the sample type. Testing is either screening or confirmatory depending on the sample as well as the purpose of testing. Screening methods are usually rapid and give qualitative detection in most cases. However, they are not specific, which means that a positive result should be considered as "presumptive positive" and must be followed by confirmatory testing, which has greater specificity.

There are several methods used for measurement and the most common ones are discussed here.

O. Z. Ismail · V. T. Chetty (✉)
Pathology and Molecular Medicine, McMaster University, Hamilton, ON, Canada
e-mail: olaismail@medportal.ca; chetty@hhsc.ca

© Springer Nature Switzerland AG 2022
R. Valani (ed.), *Cannabis Use in Medicine*,
https://doi.org/10.1007/978-3-031-12722-9_5

Analytical Techniques

- Various analytical techniques are available to determine the concentration of a compound. The analytical method depends on the chemical compound, the specimen from which the compound is being analyzed, and the physical characteristics of the substrate.
- For cannabinoids and metabolites, the following are the most commonly employed techniques used for analysis:

Immunoassay

- An immunoassay method is typically used to screen for potential cannabinoid use in the workplace, roadside drug testing, and clinical specimens [1].
- These assays rely on the reaction between antigen in the sample (THC or its metabolite) and antibody specific for THC or its metabolite.
- With immunoassays, the results are qualitative indicating that they are positive if THC or its metabolite in the sample is above a certain cut-off levels.
    - The cut-off level developed by the department of health and human services in the US for workplace testing is around 50 ng/mL. This threshold is selected to avoid false-positive results [2, 3].
    - Hence, a negative screen by immunoassay doesn't mean the absence of THC or its metabolite in the sample but instead indicates that the level in the sample is below the cut-off level of the assay. A presumptive positive sample by immunoassay should be confirmed by one of the quantitative methods described below.
- Advantages
    - These assays are inexpensive and easy to perform.
    - The turn-around time is quick.
    - Urine and oral fluid are the most common type of specimens used in these assays, and they detect THC-COOH-glucuronide and THC, respectively [4].
- Limitations
    - Cross-reactivity with medications could yield false-positive results. Some medications implicated in false positive results include proton pump inhibitors (PPIs), nonsteroidal anti-inflammatory drugs (NSAIDs) and anti-retroviral medication (efavirenz) [5].
    - This technique will not detect synthetic cannabinoids, which are human-made chemicals that bind to the cannabinoid receptors [1, 2, 6]. See Chap. 3.

Gas Chromatography-Mass Spectrometry (GC-MS)

- GC-MS is one of the most common analytical approaches for cannabinoid detection in a variety of matrix samples [1, 7].

- It is used for separating volatile compounds by combining the separation power of chromatography with high specificity and low detection limit of a mass spectrometer [8].
- Because cannabinoid compounds are lipophilic and GC-MS require the sample to be volatile, a variety of extractions and derivatization steps are employed to make cannabinoid compounds volatile.
- Several sample types could be used for GC-MS, including serum, plasma, urine, oral fluid and hair [4, 9].
- Advantages
  - The GC-MS method offers a quantitative measurement of THC and its metabolite in samples. It provides high sensitivity and specificity, which makes it a suitable method for confirming the presence of cannabinoid in the sample at very low concentrations [10].
- Limitations
  - Due to the significant time required to prepare the sample, this method is time-consuming, costly and requires highly trained personnel.
  - GC-MS has a slow analysis time with an average of 30 min per sample and offers a limited number of molecules that could be analyzed [11].

Liquid Chromatography-Tandem Mass Spectrometry (LC-MS/MS)

- LC-MS/MS is another analytical method that is used for confirmation of cannabinoid present in the sample and forensic investigations.
- It is similar to GC-MS, where two analytical techniques are coupled together.
  - Liquid chromatography involves a liquid mobile phase and a solid stationary phase. If the solid stationary phase consists of small-diameter particles, it is known as high-performance liquid chromatography.
- Various sample types could be used in the analysis by LC-MS/MS, such as serum, plasma, whole blood, urine, hair, oral fluid and breath [9].
- This method offers a higher specificity for the quantification of target compounds, especially in a complex matrix.
  - The resulting mass to charge ratios (m/z) for both the precursor and product ions are compared to expected chemical spectra for the target compounds for identification and also compared to internal standard for quantifications [8, 9, 12].
- Advantages
  - LC-MS/MS offers an advantage of rapid and straightforward sample preparation that doesn't require extensive sample derivatization as in GC-MS.
  - The LC-MS/MS is an efficient technique that allows for the simultaneous quantification of the different cannabinoid metabolites (free and conjugated forms) in a single analytical run.
- Limitation
  - The limitation for the use of LC-MS/MS in a clinical lab is high capital expenditure for the equipment and high operating costs.

Sample Types and Analytes Measured

- There are several sample types used in the clinical setting for measuring THC and its various metabolites.
- Other cannabis constituents that are often present in some sample type include cannabigerol (CBG), cannabinol (CBN) and tetrahydrocannabivarin (THCV), which lack psychoactive effects but indicate recent cannabis use if detected in the sample [1, 13]

Urine Sample

- The primary urinary cannabinoid that is often detected is carboxy-THC-glucuronide (THCCOOH-glucuronide), which is detected either directly by immunoassay or after hydrolysis steps to form carboxy-THC (THC-COOH) by GC-MS or LC-MS/MS [9, 14].
- Several studies indicated that comparing the ratio of THC-COOH glucuronide to creatinine for 2 paired specimens (collected at least 48 h apart) increased the accuracy of predicting new marijuana use, which is useful in the setting of anti-doping, law enforcement, and military cases [1].
- All of the other markers, such as, Δ9-THC, 11-OH-THC, CBD, or CBN were unmeasurable in urine in several studies [14, 15].

Blood

- Whole blood and plasma are common sample types where THC, 11-OH-THC, THC-COOH and in some cases CBG, CBN and THCV are detected.
- Using LC-MS/MS, several studies have shown that detection of CBG and CBN in blood sample indicate recent cannabis use [4, 13].
- When comparing the different routes of marijuana intake on the level of THC and its metabolite detection in whole blood, a higher sensitivity of detecting THC was observed in vaporized cannabis [1, 4].

Oral Fluid (Saliva)

- The current point of care testing assay used (or in the process of approval) at the workplace, pain management centers and roadside testing favours the use of oral fluid (saliva) as sample of choice.
- Oral fluid represents a simple, non-invasive testing matrix that is readily available. Both THC and THC-COOH are the primary cannabinoids detected in these samples.
- If using LC-MS/MS, one could also identify other cannabinoids such as THCV, CBD, and CBG for both frequent and occasional users [1, 4, 16].
- Oral fluid offers greater sensitivity in detecting THC when cannabis was smoked.

Hair

- Measuring THC-COOH in hair samples using LC-MS/MS reflects long term daily consumption of cannabis [17].
- A potential problem with hair sample measurement is the contamination of the sample by environmental cannabis smoke (secondary exposure), which could alter the results [18].

Sweat

- Measurement of THC using weekly sweat patch could be used in the criminal justice program to determine weekly cannabis use, but more research needs to be done in this area [19].

New and Upcoming Chemical Detection Method Using Breath Sample

- There is growing evidence of the ability to detect THC in exhaled breath samples following acute marijuana intake (anywhere from 30 min up to 3 h) [20, 21].
- This non-invasive sample type that reflects recent cannabinoid exposure attracted several companies to develop a portable method of measuring it.
- There are two companies currently working on developing a portable breath analyzer to measure THC, Hound lab Inc. (in US) and Cannabix technologies Inc. (in Canada).
  - The Hound lab module of a portable breath analyzer collects 5 L of breath at rate of 5 L/min. The breath will pass through a saliva trap, which measures the level of alpha-amylase in the sample. If alpha-amylase is positive, the sample would be rejected due to contamination with saliva. THC is then isolated from the sample and subsequently will react with fluorescently labelled diazonium salt made from Rhodamine 123 precursor forming a complex with emission measured at 590 nm .

Challenges and Limitations

- Even with the advances of different measuring methods for detecting cannabinoids, one should not underestimate the challenges of determining the level of cannabinoid that is associated with cognitive impairment.
- Some studies suggest that THC blood concentration of more than 5 ng/mL is associated with substantial cognitive and psychomotor impairment.
  - The drugs and driving Committee's current cut-off concentration for THC is 25 ng/mL, which someone could argue could be an average level for chronic users and wouldn't produce any impairment.

- The lack of concordance between the concentration of analyte and cognitive status is explained by the difference seen between acute and chronic users, who might have different effects at the same blood THC level. For cannabinoid, the concentration-effect curves are not linear or sigmoidal but rather a counterclockwise hysteresis curves.
- There are usually two phases that occur after cannabis intake. The first phase is the absorption phase, where the psychoactive effects are low, which is followed by much higher effects during the distribution phase of cannabinoid to the brain. Therefore, chronic cannabinoid users must achieve a more elevated blood THC concentrations to reach the same cognitive impairment level as occasional users. Hence, this limits the clinical value of using a single cut off level for interpreting cannabinoid levels in the context of cognitive and psychomotor effects.

References

1. Huestis MA, Smith ML. Cannabinoid markers in biological fluids and tissues: revealing intake. Trends Mol Med. 2018;24(2):156–72.
2. Moeller KE, et al. Clinical interpretation of urine drug tests: what clinicians need to know about urine drug screens. Mayo Clin Proc. 2017;92(5):774–96.
3. Macdonald S, et al. Testing for cannabis in the work-place: a review of the evidence. Addiction. 2010;105(3):408–16.
4. Pacifici R, et al. THC and CBD concentrations in blood, oral fluid and urine following a single and repeated administration of "light cannabis". Clin Chem Lab Med. 2020;58(5):682–9.
5. Rollins DE, Jennison TA, Jones G. Investigation of interference by nonsteroidal anti-inflammatory drugs in urine tests for abused drugs. Clin Chem. 1990;36(4):602–6.
6. Diao X, Huestis MA. Approaches, challenges, and advances in metabolism of new synthetic cannabinoids and identification of optimal urinary marker metabolites. Clin Pharmacol Ther. 2017;101(2):239–53.
7. Musshoff F, Madea B. Review of biologic matrices (urine, blood, hair) as indicators of recent or ongoing cannabis use. Ther Drug Monit. 2006;28(2):155–63.
8. Maurer HH. Mass spectrometry for research and application in therapeutic drug monitoring or clinical and forensic toxicology. Ther Drug Monit. 2018;40(4):389–93.
9. Leghissa A, Hildenbrand ZL, Schug KA. A review of methods for the chemical characterization of cannabis natural products. J Sep Sci. 2018;41(1):398–415.
10. Day D, et al. Detection of THCA in oral fluid by GC-MS-MS. J Anal Toxicol. 2006;30(9):645–50.
11. Andrenyak DM, et al. Determination of Δ-9-tetrahydrocannabinol (THC), 11-hydroxy-THC, 11-nor-9-carboxy-THC and cannabidiol in human plasma using gas chromatography-tandem mass spectrometry. J Anal Toxicol. 2017;41(4):277–88.
12. Tiscione NB, et al. An efficient, robust method for the determination of cannabinoids in whole blood by LC-MS-MS. J Anal Toxicol. 2016;40(8):639–48.
13. Newmeyer MN, et al. Free and glucuronide whole blood cannabinoids' pharmacokinetics after controlled smoked, vaporized, and oral cannabis administration in frequent and occasional cannabis users: identification of recent cannabis intake. Clin Chem. 2016;62(12):1579–92.
14. Huestis MA, Cone EJ. Differentiating new marijuana use from residual drug excretion in occasional marijuana users. J Anal Toxicol. 1998;22(6):445–54.
15. Fabritius M, et al. THCCOOH concentrations in whole blood: are they useful in discriminating occasional from heavy smokers? Drug Test Anal. 2014;6(1–2):155–63.

16. Swortwood MJ, et al. Cannabinoid disposition in oral fluid after controlled smoked, vaporized, and oral cannabis administration. Drug Test Anal. 2017;9(6):905–15.
17. Huestis MA, et al. Cannabinoid concentrations in hair from documented cannabis users. Forensic Sci Int. 2007;169(2–3):129–36.
18. Berthet A, et al. A systematic review of passive exposure to cannabis. Forensic Sci Int. 2016;269:97–112.
19. Huestis MA, et al. Excretion of Delta9-tetrahydrocannabinol in sweat. Forensic Sci Int. 2008;174(2–3):173–7.
20. Coucke L, et al. $\Delta(9)$-tetrahydrocannabinol concentrations in exhaled breath and physiological effects following cannabis intake—a pilot study using illicit cannabis. Clin Biochem. 2016;49(13–14):1072–7.
21. Himes SK, et al. Cannabinoids in exhaled breath following controlled administration of smoked cannabis. Clin Chem. 2013;59(12):1780–9.

The Role of Genetics in the Use of Cannabis

6

Michelle Di Risio and Prakash Gowd

Introduction

There are over 100 cannabinoids that have been isolated from the cannabis plant. Of the many compounds, the two most studied are Δ9-Tetrahydrocannabinol (THC) and Cannabidiol (CBD). Each of these has a distinct affinity for receptors in the body (see Chap. 3).

THC is the major psychotropic constituent of cannabis drugs isolated from the plant *Cannabis sativa*. It is a psychoactive substance that acts by binding to the Cannabinoid Type 1 (CB_1) receptor in the brain [1]. THC has been shown to have therapeutic value as an analgesic, muscle-relaxant, appetite-stimulant, antiemetic, neuroprotective, and anti-cancer drug [2]. Individual users have different reactions to cannabis depending on how their bodies process the chemical constituents. Common adverse effects of THC include sedation, anxiety, dizziness, dysphoria, ataxia, changes in visual perception, dry mouth, altered sense of time and red eyes.

CBD, on the other hand, has a greater affinity to the CB_2 receptor and interacts differently with the nervous system without eliciting any psychoactive effect. CBD provides potential health benefits as an antioxidant, antiemetic, anticonvulsant and anti-tumoral agent. CBD can also counteract the psycho-activity of THC and acts as a natural antidepressant and neuroprotective agent. Common side effects of CBD include dizziness, nausea vomiting, diarrhea, and drowsiness [3].

While THC and CBD are both known to have therapeutic value, specific genetic polymorphisms have been demonstrated to (a) affect an individual's rate of

M. Di Risio (✉)
Medical Business Development, Lobo Genetics Inc., Toronto, ON, Canada
e-mail: michelle.dirisio@lobogene.com

P. Gowd
Lobo Genetics Inc., Toronto, ON, Canada

© Springer Nature Switzerland AG 2022
R. Valani (ed.), *Cannabis Use in Medicine*,
https://doi.org/10.1007/978-3-031-12722-9_6

cannabinoid metabolism; (b) predispose users to the development of cannabis-induced psychotic disorders; and (c) cause negative effects on neurocognition. THC and CBD genetic testing offers a solution to identify inter-individual variations to ensure optimal cannabinoid drug effects, while minimizing short- and long-term negative consequences.

Cytochrome P450 Enzymes

- The cytochrome P450 (CYP) enzymes are responsible for the phase I metabolism of several endogenous and exogenous substrates, and oxidize more than 90% of current therapeutic drugs.
- The CYP gene encodes the cytochrome P450 class of metabolic enzymes. The majority of the CYP enzymes are found in the liver.
- Genetically determined differences in drug-metabolizing enzymatic activity can lead to inter-individual variability in drug responses, resulting in altered efficacy and toxicity in patients.
    - CYP genetic variations produce inter-individual differences in THC and CBD metabolism. It is therefore important to consider THC and CBD inter-individual metabolic variability when determining type of cannabis strain to be used and dosing for a particular patient.

Cytochrome P450 Enzyme: CYP2C9

- CYP2C9 is major enzyme responsible for the metabolism of THC.
- CYP2C9 metabolizes THC by hydroxylating the cannabinoid to form 11-hydroxy-Δ9 THC (Δ9-THC-OH, or simply THC-OH).

$$\text{THC} \xrightarrow{P450\,system} \text{THC}-\text{OH} \xrightarrow{CYP2C9} \text{THC}-\text{COOH} \xrightarrow{Glucoronidation} \text{THC}-\text{COOH}-\text{glucoronide}$$

    - THC-OH is an active metabolite that has significant psychotropic activity.
    - THC-OH is further oxidized by CYP2C9 to form the carboxylic acid metabolite 11-nor-9-carboxy-Δ9 THC (THC-COOH).
    - THC-COOH is a pharmacologically inactive metabolite as it can no longer tightly bind to the CB_1 receptor [2].
- Research studies have shown that 15–20% of the population has difficulty metabolizing THC from its psychoactive to non-psychoactive form.
    - These individuals can be more sensitive to THC's psychoactive effects compared to those with normal metabolism.
    - The THC metabolic variation is caused by CYP2C9 polymorphisms.

CYP2C9 Polymorphisms

- CYP2C9 polymorphisms affect the rate of THC metabolism and therefore can impact individual responses to THC.
  - Specific CYP2C9 gene variants can help to determine THC's therapeutic effect and mitigate any possible drug side effects.
  - A patient's THC response can be better understood by identifying their CYP2C9 genotype. This can help tailor dosing and counsel around side effects.
- The CYP2C9*1 allele is considered the wild-type allele or "normal" allele, with "normal" enzyme activity.
  - The CYP2C9*2 and CYP2C9*3 genotypes are other well-known genetic variants in clinical pharmacokinetic studies and are the most common loss-of-function alleles [4].
  - Carriers of these two alleles (CYP2C9*2 and CYP2C9*3) have reduced CYP2C9 enzyme activity.
- Individuals with the CYP2C9*3 genetic variation convert THC into its non-psychoactive form much slower than the average person.
  - Those with two copies of the CYP2C9*3 allele (homozygotes) are considered "very slow metabolizers", while those with one copy of the allele (heterozygotes) are called "slow metabolizers" (see Table 6.1).

Clinical Significance

- Clinical research has shown individuals with slower THC metabolism due to the CYP2C9*3 gene experienced psychoactive THC levels that are 200–300% higher than the average individual. Furthermore, CYP2C9*3 carriers displayed a trend toward increased sedation and duration of sedation following administration of THC [2, 5].
  - These effects may be due to the low catalyzed formation of the inactive metabolite, THC-COOH, measured in CYP2C9*3 carriers.
- The slower metabolism of THC can increase the likelihood of negative side effects, including extreme anxiety, hallucinations, paranoia, rapid heart rate and panic attacks [6]. Therefore, careful titration is warranted in these individuals.

Table 6.1 CYP2C9 phenotypes and genotypes

Metabolizer phenotype	Enzyme activity	Metabolizer genotype
Extensive metabolizer	Normal	*1/*1
Slow metabolizer	Reduced	*1/*3
Very slow metabolizer	Very reduced	*3/*3

- Prevalence of CYP2C9*3 allele variant combined with a trend towards increasingly potent THC in cannabis could well explain the recent increase in cannabis poisoning.
  - The Canadian Institute for Health Information report showed that the number of hospital visits due to cannabis poisoning has significantly increased in the past 5 years [7].
- Evidence has also shown that there is a relationship between blood THC concentration and performance in controlled driving-simulation studies.
  - In an accident culpability analysis, individuals testing positive for THC, particularly those with higher blood levels, were 3–7 times more likely to be responsible for a motor-vehicle accident compared to a person who had not used cannabis [8].
  - A meta-analysis of nine case-control and culpability studies found that recent cannabis use doubled the risk of a car crash [9].
  - Further studies are needed to better understand the risks associated with THC use and genetic susceptibility.

Practical Implications to CYP2C9 Genetic Testing

- Genetic testing for the 2C9 metabolic status is beneficial to determine if patients are slow metabolizers of THC.
- Genetic screening, specifically for the CYP2C9*3 carriers, can enable the safer use of cannabis.
- Health care professionals (HCPs) can prescribe lower THC strains to optimize individual cannabis dosing with less side effects.

AKT1 Protein Kinase

- The AKT1 gene encodes a protein kinase that is involved in multiple cellular functions, such as cell proliferation, cell survival and transcription.
- AKT1 has a role in regulating neuronal cell size and survival and is also a key signaling molecule downstream of the dopamine D_2 (DRD2) receptor.
  - In vitro and in vivo studies have confirmed cannabinoids can stimulate the AKT1 pathway via the CB_1 and CB_2 receptors [10].
  - AKT1 kinase functionality may result in enhanced responses to DRD2 receptor stimulation.
- Dopamine is released in the brain when THC binds to the CB_1 receptors.
  - THC continuously alters an individual's brain chemistry and dopamine levels.
  - The THC-mediated increase in dopamine release may be further exacerbated by decreased AKT1 kinase functionality.
  - Elevated levels of dopamine are known to have a role in the development of psychotic symptoms [11].

- When dopamine metabolism in the brain is disrupted, excess dopamine levels can lead to short term anxiety, panic, insomnia, delirium and paranoia.
  Longer-term disruption can lead to psychosis and schizophrenia.
  Habitual chronic users can develop cannabis-induced psychosis and schizophrenia.

AKT1 Polymorphisms

- Genetic factors may confer vulnerability to psychosis outcomes following exposure to cannabis, i.e., gene-environment interaction.
- AKT1 gene polymorphism has been implicated in playing a role in moderating the association between cannabis use and psychotic disorders, such as psychosis and schizophrenia [10].
- Individuals who carry the AKT1 "C" allele have reduced AKT1 kinase functionality.
- A decrease in AKT1 functionality can influence an individual's risk of experiencing short and long-term adverse mental effects with cannabis use.

Clinical Significance

- Studies have shown the importance of AKT1 genotypes being associated with a psychotic disorder [12, 13].
  - Carriers of the AKT1 C/C genotype with a history of cannabis use had a twofold increase in the risk of being diagnosed with a psychotic disorder when compared to T/T carriers [12, 13].
  - Daily cannabis users with AKT1 C/C genotype had a sevenfold increase in the risk of being diagnosed with a psychotic disorder when compared to T/T carriers [13].
  - AKT1 C allele (C/C or C/T) were significantly associated with acute psychotic symptoms after smoking cannabis [11].

Practical Implications to AKT1 Genetic Testing

- Genetic testing for AKT1 C allele carriers can identity individuals who have a genetic predisposition that exposes them to a higher risk of cannabis-induced psychosis.
- Genetic screening, specifically for the AKT1 C carriers, can enable the safer use of cannabis.
- Health care professionals (HCPs) can prescribe lower THC strains to optimize individual cannabis dosing with less mental health side effects.

Catechol-*O*-Methyltransferase (COMT) Enzyme

- The COMT gene encodes the Catechol-*O*-Methyl-Transferase (COMT) enzyme that plays an important role in the degradation of dopamine in the brain.
- COMT breaks down dopamine mostly in the part of the brain responsible for higher cognitive or executive function (prefrontal cortex).
- Lower dopamine levels can affect adversely neurocognitive function.

COMT Polymorphisms

- Genetic factors may explain the difference in an individual's sensitivity to the acute neurocognitive impairment-inducing effects of cannabis.
- COMT gene polymorphisms is linked to differences in enzyme activity.
    - A COMT gene functional polymorphism is a G to A missense mutation that generates a valine (Val) to methionine (Met) substitution at codon 158 (Val158Met).
    - This codon substitution produces lower enzymatic activity and a slower breakdown of dopamine.
        Met/ Met homozygotes have the highest levels of dopamine (due to low COMT enzyme activity)
        Val/Val homozygotes have the lowest levels of dopamine (due to high COMT enzyme activity); and heterozygotes intermediate levels (due to intermediate COMT enzyme activity).
- It has been suggested that the COMT Val158Met polymorphism influences acute effects of THC on various neurocognitive functions following cannabis exposure [14].

Clinical Significance

- Clinical studies have demonstrated COMT Val/Met polymorphism may influence the acute effects of THC on various neurocognitive functions [14–17].
    - COMT Val allele carriers were more sensitive to THC-induced working memory and attention impairments compared to Met carriers [15].
    - Cannabis users with the COMT Val/Val genotype exhibited lower accuracy of sustained attention and were less prone to respond versus non-cannabis users carrying the same genotype [16].
    - COMT Val allele carriers committed more monitoring/shifting errors than cannabis users who are homozygous for the COMT Met genotype [17].

Practical Implications to COMT Genetic Testing

- Genetic testing for COMT Val allele carriers can identity individuals who have a genetic predisposition that exposes them to a higher risk of acute neurocognitive impairment in combination with cannabis use [5, 18, 19].
- Genetic screening, specifically for carriers of the COMT Val carriers, can enable the safer use of cannabis.
- Health care professionals (HCPs) can prescribe lower THC strains to optimize individual cannabis dosing with less acute neurocognitive adverse effects.

Cytochrome P450 Enzyme: CYP2C19

- The CYP2C19 enzyme metabolizes CBD which undergoes extensive hydroxylation at multiple sites and further oxidations results in several different CBD metabolites.
  - The major metabolites of CBD are hydroxylated 7-OH-CBD derivatives; however, less is known about the CBD metabolites compared to THC [20].
- There are inter-individual differences in the expression and function of the CYP2C19 enzyme which can considerably affect the pharmacokinetics of CBD and its metabolites.
- These differences may be relevant in the therapeutic action and possible adverse effects of CBD-containing preparations.
- These functional variations are caused by CYP2C19 polymorphisms.

CYP2C19 Polymorphisms

- The CYP2C19*1 allele is considered the wild-type allele or "normal" allele, with "normal" enzyme activity.
  - CYP2C19*2 and CYP2C19*3 alleles are the most common loss-of-function alleles.
      Carriers of these two alleles have reduced CYP2C19 enzyme activity.
  - The CYP2C19*17 allele in another polymorph and is a gain-of-function allele.
      Carriers of this allele have increased CYPC219 enzyme activity.
  - There are alleles other than *2, *3 and *17, however these alleles are carried by less than 0.5% of the individuals in most ethnic groups (see Table 6.2).

Table 6.2 CYP2C19 phenotypes and genotypes

Metabolizer phenotype	Enzyme activity	Metabolizer genotype
Extensive metabolizer	Normal	*1/*1
Intermediate metabolizer	Reduced	*1/*2, *1/*3, *2/*17, *3/*17
Poor metabolizer	Very reduced	*2/*2, *3/*3, *2/*3
Ultra-rapid metabolizer	Increased	*1/*17, *17/*17

Clinical Significance

- It is highly likely that individuals with a poor metabolizer CYP2C19 phenotype who take CBD have higher levels of CBD versus individuals with an extensive metabolizer phenotype, although no large study has been generated in the literature to-date.
- Furthermore, it is well known that there are several common drugs that are metabolized by the CYP2C19 enzyme.
- Drugs that decrease CYP2C19 activity (inhibitors) or increase activity (inducers) need to be monitored carefully in patients who are taking CBD [4].

Practical Implications to CYP2C19 Genetic Testing

- Genetic testing for the 2C19 metabolic status is beneficial to determine different CYP2C19 enzyme metabolic activity.
- Genetic screening, specifically for the CYP2C19*2, *3 and *17 carriers, can enable the safer use of cannabis.
- Health care professionals (HCPs) can tailor CBD dosing to optimize clinical effects and potentially mitigate adverse drug reactions with other therapies.

Conclusion

- Genetic testing for CYP2C9, AKT1 and COMT polymorphisms supports safer cannabis dosing by assessing THC metabolism status (normal, slow and very slow metabolizers) as well as identifying individuals with short- and long-term negative risk factors that predispose them to cannabis-induced psychosis and acute detrimental neurocognitive impairments with medical cannabis consumption.
- Genetic testing for CYP2C19 polymorphisms supports safer cannabis dosing by assessing CBD metabolism status (extensive, intermediate, poor or ultra-rapid metabolizer) that enables healthcare professionals to optimize appropriate therapeutic dosing and minimize adverse reactions in patients who are using medical cannabis alone or in combination with other therapies.

References

1. Bland TM, et al. CYP2C-catalyzed delta9-tetrahydrocannabinol metabolism: kinetics, pharmacogenetics and interaction with phenytoin. Biochem Pharmacol. 2005;70(7):1096–103.
2. Sachse-Seeboth C, et al. Interindividual variation in the pharmacokinetics of Delta9-tetrahydrocannabinol as related to genetic polymorphisms in CYP2C9. Clin Pharmacol Ther. 2009;85(3):273–6.

3. MacCallum CA, Russo EB. Practical considerations in medical cannabis administration and dosing. Eur J Intern Med. 2018;49:12–9.
4. Hirota T, Eguchi S, Ieiri I. Impact of genetic polymorphisms in CYP2C9 and CYP2C19 on the pharmacokinetics of clinically used drugs. Drug Metab Pharmacokinet. 2013;28(1):28–37.
5. Government of Canada. Information for health care professionals. Cannabis (marihuana, marijuana) and the cannabinoids. 2018. https://www.canada.ca/content/dam/hc-sc/documents/services/drugs-medication/cannabis/information-medical-practitioners/information-health-care-professionals-cannabis-cannabinoids-eng.pdf.
6. National Institute of Drug Abuse. Marijuana. 2018. https://d14rmgtrwzf5a.cloudfront.net/sites/default/files/1380-marijuana.pdf.
7. Nicholson. Spike in cannabis overdoses blamed on potent edibles, poor public education. 2018. https://www.cbc.ca/news/health/cannabis-overdose-legalization-edibles-public-education-1.4800118.
8. Volkow ND, Compton WM, Weiss SR. Adverse health effects of marijuana use. N Engl J Med. 2014;371(9):879.
9. Hall W. What has research over the past two decades revealed about the adverse health effects of recreational cannabis use? Addiction. 2015;110(1):19–35.
10. Radhakrishnan R, Wilkinson ST, D'Souza DC. Gone to pot—a review of the association between cannabis and psychosis. Front Psych. 2014;5:54.
11. Morgan CJ, et al. AKT1 genotype moderates the acute psychotomimetic effects of naturalistically smoked cannabis in young cannabis smokers. Transl Psychiatry. 2016;6(2):e738.
12. van Winkel R. Family-based analysis of genetic variation underlying psychosis-inducing effects of cannabis: sibling analysis and proband follow-up. Arch Gen Psychiatry. 2011;68(2):148–57.
13. Di Forti M, et al. Confirmation that the AKT1 (rs2494732) genotype influences the risk of psychosis in cannabis users. Biol Psychiatry. 2012;72(10):811–6.
14. Cosker E, et al. The effect of interactions between genetics and cannabis use on neurocognition. A review. Prog Neuropsychopharmacol Biol Psychiatry. 2018;82:95–106.
15. Henquet C, et al. An experimental study of catechol-o-methyltransferase Val158Met moderation of delta-9-tetrahydrocannabinol-induced effects on psychosis and cognition. Neuropsychopharmacology. 2006;31(12):2748–57.
16. Tunbridge EM, et al. Genetic moderation of the effects of cannabis: catechol-O-methyltransferase (COMT) affects the impact of Δ9-tetrahydrocannabinol (THC) on working memory performance but not on the occurrence of psychotic experiences. J Psychopharmacol. 2015;29(11):1146–51.
17. Verdejo-García A, et al. COMT val158met and 5-HTTLPR genetic polymorphisms moderate executive control in cannabis users. Neuropsychopharmacology. 2013;38(8):1598–606.
18. Schiffman J, et al. Attitudes towards cannabis use and genetic testing for schizophrenia. Early Interv Psychiatry. 2016;10(3):220–6.
19. National Institute of Drug Abuse. Is there a link between marijuana use and psychiatric disorders? 2020. https://www.drugabuse.gov/publications/research-reports/marijuana/there-link-between-marijuana-use-psychiatric-disorders.
20. Ujváry I, Hanuš L. Human metabolites of cannabidiol: a review on their formation, biological activity, and relevance in therapy. Cannabis Cannabinoid Res. 2016;1(1):90–101.

Evidence Based Reviews on Cannabis Therapy

7

Tanmay Sharma and Suneel Upadhye

Introduction

Modern medical practice is increasingly guided by evidence from empirical research. This approach to the practice of medicine, known as evidence-based medicine (EBM), which first came to prominence only about three decades ago, has transformed the practice of medicine in recent times and has been considered one of the most significant innovations in healthcare and medicine. In a seminal paper by EBM pioneer David Sackett, EBM was described as "the conscientious, explicit and judicious use of the current best evidence in making decisions about the care of individual patients" [1]. Before the establishment and mainstream recognition of the EBM paradigm, clinical decision making commonly relied on expert opinion, physiological reasoning and some unstructured use of evidence—an approach sometimes referred to as "eminence-based medicine." However, a growing number of inconsistencies between research evidence and expert recommendations eventually prompted a shift towards a system of evidence-based decision-making. Today, clinicians globally are encouraged to use the best available evidence from systematic research and integrate it with individual clinical expertise and patient preferences to guide their practice [2].

T. Sharma
Schulich School of Medicine and Dentistry, Western University, London, ON, Canada
e-mail: tsharma2024@meds.uwo.ca

S. Upadhye (✉)
Emergency Medicine/Health Research Methods, Evidence and Impact (HEI), McMaster University, Hamilton, ON, Canada

R. Valani (ed.), *Cannabis Use in Medicine*,
https://doi.org/10.1007/978-3-031-12722-9_7

Types of Study Designs in Health Research

- Clinical research questions can be answered through different types of studies that fundamentally differ in their design. Each study design has different methodological strengths and weaknesses.
- Studies are typically categorized as being either descriptive or analytic.
- Descriptive studies simply examine the overall characteristics and distribution of particular factors in the sample without quantifying relationships between those factors. Some common types of descriptive studies are:
  - *Case Report or Case Series*—Anecdotal description of either a single patient or a series of patients that is normally retrospective in nature and does not involve a comparison group
  - *Cross-Sectional Survey (Descriptive)*—A 'snapshot' description of the distribution and frequency of exposures or outcomes in a defined population at one particular point in time
- While descriptive studies are good summary indicators and provide the 'lay of the land' in terms of the overall distribution of outcomes and exposures of interest, they may not always provide quantitative data on the association between these factors
- Analytic studies typically investigate the association between exposures or interventions and outcomes through a comparative analysis of two or more groups. Some common types of analytic studies are:
  - *Randomized Controlled Trials (RCTs)*—a comparative investigation wherein participants are randomly allocated to either an intervention or control/comparison group.

    The randomization of participants plays a critical role in controlling bias as it works towards creating an equal distribution of known and unknown prognostic factors within the study groups.

    This aspect of random participant allocation is what fundamentally separates RCTs from other types of analytic studies and makes it the gold standard study design among primary studies in therapy (or other interventional) research. In comparison to RCTs, non-randomized studies may present a relatively higher (or different) risk of bias due to prognostic imbalance between groups.
  - *Cohort Study*—a type of non-randomized study that compares the distribution of outcomes (e.g., lung cancer incidence) between two or more groups based on differences in their exposure status (e.g., smoker vs. non-smoker). Hence, in cohort studies, participants are particularly followed from exposure to outcome.

    Given that exposure is identified before the outcome, these studies can assess causal associations, and are especially valuable in addressing research questions about harms and prognosis. However, they are resource-intensive and are not optimal for studying rare outcomes.
  - *Case-Control Study*—a type of non-randomized, retrospective study that examines the distribution of exposures between two or more groups based on differences in their disease status (i.e., cases vs. controls).

In case-control studies, participants are followed from outcome to exposure. These studies are relatively quick and inexpensive, and are optimal when rare outcomes are being studied, or when the induction period from exposure to disease is long. However, they cannot be used to assess outcome incidence and hence are not optimal for inferring causality.

- *Cross-Sectional Study (Analytic)*—a type of 'snapshot' study, which unlike longitudinal studies, examines the relationship between exposures and outcomes of interest at only one particular point in time (i.e., involves no follow up over time). This type of study is best suited to assess the prevalence of a disease or risk factors at a particular time, and for examining accuracy of diagnostic tests.

- *Systematic Review*—this is a type of second-order research study that uses a structured and methodical approach to consolidate data from all relevant primary studies, undergo quality assessments of included trials, and may provide a pooled summary of the results to answer a research question.

    Such reviews may also include Scoping reviews (no quality assessments completed), and individual patient data (IPD) reviews (where the unit of analysis are individual patients, not studies).

    These different reviews are well described using PRISMA (Preferred Reporting Items for Systematic Reviews and Meta-Analyses) materials (accessible at: http://www.prisma-statement.org).

- *Meta-Analysis*—an extension of a systematic review that involves the use of statistical techniques to quantitatively pool results from all relevant primary studies and determine a single pooled effect estimate based on the results obtained from the individual studies.

    Alternative meta-analyses may include individual patient data (IPD) studies, and network meta-analyses to allow for indirect comparisons between interventions (see PRISMA materials).

Hierarchy of Evidence: The Traditional Model

- One of the fundamental doctrines of EBM indicates that not all evidence is the same. The different types of studies vary in methodological merit (and associated risks of bias), and a hierarchy of evidence exists.
- Traditionally, the quality of research evidence in studies has been largely judged based on the design of the studies. This is manifested in the traditional evidence pyramid model which ranks the various types of studies based on their internal validity. See Fig. 7.1.
- In the pyramid model, RCTs are placed at the top of the primary research hierarchy, followed by observational study designs such as cohort studies and case-control studies in the middle, and unsystematic clinical observation reports such as case series and case reports at the bottom.
- Since synthesized evidence usually is more informative/useful than primary studies, they have generally been considered higher sources of evidence. Hence,

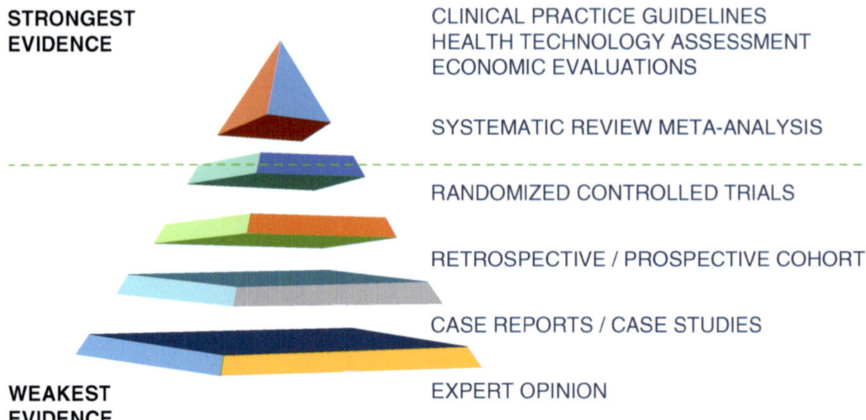

Fig. 7.1 Traditional hierarchies of medical evidence

in the pyramid model, for the study of therapies/interventions, systematic reviews and meta-analyses of primary studies are placed at a higher level of the pyramid hierarchy.

- Amalgamating synthesized clinical evidence with clinical judgement and patient/stakeholder inputs leads to the production of clinical practice guidelines, which usually generate specific practice recommendations for unique questions.
  - These recommendations should incorporate clinical evidence evaluations, patient inputs (values/preferences, prioritization, equity considerations), resource implications and feasibility/acceptability assessments, using valid frameworks such as the Grading of Recommendations, Assessment, Development and Evaluations (GRADE) Evidence-to-Decision framework [2]
  - Such guidelines should be methodologically sound and result in trustworthy recommendations that are readily adaptable/adoptable into clinical practice.
- Finally, all the preceding levels of evidence can be integrated, with specific resource costing measures, to create formal economic evaluations and health technology assessments, that may be of interest to clinicians, institutions and health services payors (e.g., government, insurance, health maintenance organizations).
  - Specific methodological standards are required to produce proper cost-effectiveness analyses (CEA), cost-utility analyses (CUA), and cost-benefit analyses (CBA) [3].
- It is important to note that the hierarchy varies depending on the area of research as some study designs are better suited than others within certain contexts.
  - Readers should appreciate that "lower" studies with strong methodological quality/rigor may provide more useful evidence than "higher" studies with weaker quality.
  - An appreciation of methodological standards for different study designs, in order to identify potential biases that may skew/distort study results.

The Revised Model

- While the traditional hierarchy can serve as a starting point indicator, in recent years this model has been considered too simplistic for drawing conclusions about the quality or certainty of evidence. For example, a methodologically strong and directly relevant randomized trial (compared to a weaker/poor quality review or guideline) may provide more direct guidance to influence bedside patient care & shared decision-making, which is the ultimate application of evidence-based medicine.
- The traditional hierarchy model still holds some merit in that study design is still considered an important factor when assessing evidence quality, and homogenous unbiased systematic reviews are still considered superior sources of evidence in most cases compared to individual studies. However, it is important to note that there are additional factors that also impact the quality of evidence, as they apply to the reader. These may include local generalizability to individual care, economic (or other resource) considerations, patient values/preferences, or other equity considerations.
- The use of validated critical appraisal tools for different study designs can help readers to detect the presence, and possible magnitude, of various methodological biases that may distort the final study findings. Many such tools are available online, and some are summarized on the EQUATOR network website (Accessible at: https://www.equator-network.org).

Critical Appraisal Tools (CATs)

- One of the critical components of appraising scientific literature is assessing the likelihood or risk of bias in the results.
- Study results can be biased at several different points in the life cycle of a study due to a variety of different reasons. In primary randomized trials, for instance, results can be influenced by factors pertaining to selection of enrolled patients, balancing demographic predictors prior to intervention, blinding of interventions and outcomes assessments, analytical plans (e.g. intention-to-treat vs. per-protocol), and losses to follow-up.
  - It is important to be familiar with the potential biases that may arise with study deficiencies related to such factors.
- Various critical appraisal tools have been developed to help assess bias within studies.
  - Most of these tools are structured into domains, include guidance regarding factors to consider when appraising a study, and provide an algorithm to ultimately rate the risk of bias as either low, moderate, or high.
- Examples of some critical appraisal tools that are commonly used for evaluating risk of bias include:
  - Cochrane ROB 2 (*Randomized Intervention Studies*)
  - Cochrane ROBIN–I (*Non-randomized Intervention Studies*)

- *Newcastle-Ottawa Scale (Non-randomized Intervention Studies)*
- QUIPS (*Prognostic Studies*)
- QUADAS-2 (*Diagnostic Studies*)
- AMSTAR 2 (*Systematic Review of Intervention Studies*)
- CHEERS (*Economic Evaluations*)
- It is important to ensure that when using various CATs that they have been validated for use in different analyses (i.e., specialties, users).
  - Such tools should obey the rules of psychometric validity and reliability, in order to be useful in critical appraisal exercises [4].
- While risk of bias assessment is one of the core aspects of evidence appraisal, quality of evidence is also influenced by several other factors.
  - There are different critical appraisal frameworks available that account for these different factors and hence allow for a holistic evaluation of the quality of evidence. Some of the most used systems within this context are:
    - The GRADE approach—provides four ratings for level of confidence: high, moderate, low, and very low.
    - The USPSTF approach—provides three ratings for level of confidence: high, moderate, and low.
- These approaches are intended for evaluating overall bodies of evidence, ideally consolidated through systematic reviews, rather than individual studies
- The level of evidence model proposed by the Oxford Centre of Evidence Based Medicine (OCEMB) is also another popular tool used by practitioners as a quick "heuristic" for approximating the confidence in study results from either systematic reviews or individual studies [5].
- The most recent OCEMB 2011 levels of evidence framework provides five level of evidence ratings.
  - Unlike the GRADE and USPSTF, the OCEMB heuristic is not as comprehensive and hence does not provide a definitive judgment about quality of evidence

Clinical Practice Guidelines

- Clinical practice guidelines (CPGs) are commonly designed to help healthcare practitioners in the clinical decision-making process. They are developed after thorough consideration of all relevant research evidence (e.g. supporting systematic reviews/meta-analyses) and are intended to provide clear recommendations for action to optimize patient care.
- In moving from evidence to recommendations, CPGs represent a further level of evidence processing, as they fundamentally rest upon synthesis and critical appraisal of relevant evidence, consideration of patient values/preferences/priorities/equity considerations, and assessment of the benefits and harms (clinical, resource utilization, etc.) associated with different possible courses of action. Ideally, "Strong" CPG recommendations should be operationalized into quality improvement performance metrics for implementation in optimizing patient care.

- While different assessment frameworks are available for use by guideline developers to evaluate the quality of evidence and the strength of recommendations in guidelines, in recent years a growing number of organizations globally have supported the use of the GRADE approach as a common, systematic framework within this context [2].

Grades of Recommendations

- There are multiple frameworks available to formulate and rate CPG recommendations.
    - Some of the most commonly used frameworks include the ones proposed by the GRADE working group, the USPSTF, and the American Heart Association (AHA).
- While there are certain differences between these frameworks, all three approaches include a system to rate the credibility of evidence, and a classification system to differentiate between strong clinical recommendations, which can be applied to the majority of patients, and weak clinical recommendations, which warrant further deliberation on a case-by-case basis in light of a patient's condition and preferences.
- In recent years, the GRADE approach has been increasingly endorsed by numerous organizations internationally including the World Health Organization and the Cochrane Collaboration.
    - According to the GRADE framework, recommendations can be classified as either "strong" or "weak/conditional" based on a list of factors generated with the GRADE Evidence-to-Decision (EtD) framework.
    - While one of the contributing factors that influences the strength of recommendations is the quality of the underlying evidence, additional factors also that play a critical role include the variability in patients' values and preferences, the balance between the benefits and downsides, and uncertainty around optimal use of resources.
- It is important to note that even in areas where a large body of randomized trials and systematic reviews exist, recommendations can still be weak due to influence from the other factors mentioned above.

From Evidence to Practice

- Evidence-based medicine is fundamentally based upon the idea of using the "best" available evidence within the local context of a particular patient's situation (e.g. clinical condition, values/preferences, resource implications, etc.), and the clinicians experience and best judgments. See Fig. 7.2.
- In modern medicine, evidence-based summaries and guidelines, which usually provide actionable recommendations for practice, serve as an important interphase in the evidence to practice continuum.

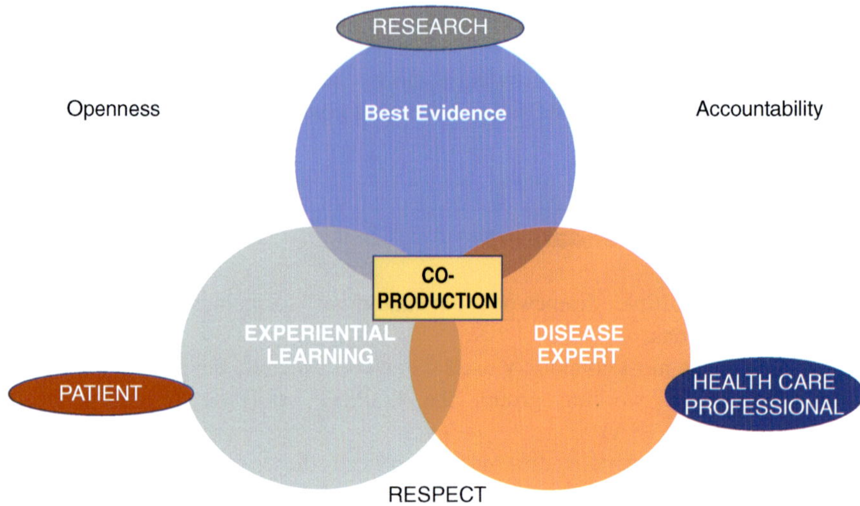

Fig. 7.2 EBM is at the centre of the best evidence, clinical practice, and patient preference

- Examples of EBM point of care tools that are commonly used by clinicians include UpToDate, DynaMed, and BMJ Best Practice.
- For healthcare professionals, the practice of EBM should ideally involve continuous assessment of evidence-based recommendations and a discussion of these recommendations with patients within the context of outcomes that are important to them.
- When using guideline recommendations, the following factors should be kept in mind:
  - If a recommendation is strong, it can be typically applied to the majority of the patients in all or almost all cases, without much review of the underlying evidence and without a detailed discussion with the patient. There may still be some rare exceptions to this case.
  - If a recommendation is weak/conditional, considering patients' values and preferences is imperative. In this case, the underlying evidence should ideally be reviewed and the potential benefits and harms of the proposed course of action should be thoroughly discussed with the patient so that the decisions are ultimately based on both the best evidence and patients' values and preferences.

Patient Engagement in Cannabis Research

- It is increasingly important in clinical research to engage patient stakeholders in the research development process to ensure patient-focused outcomes [6–8].
- Meaningful participation of patient and public involvement (PPI) in the research process can help focus research activities, increase generalizability of results, enhance likelihood of publication, and receiving grant funding.

- The evolving field of Experience Based Co-Design (EBCD) describes the meaningful recruitment, training and participation of patient partners in the research process [9].
- Patients with lived experiences enrich the research development (co-production) and knowledge translation process, in order to achieve optimal outcomes for future patient recipients of new clinical care recommendations.

Conclusions

Evidence-based medicine at the bedside is the intersection of clinicians using their experience, expertise, and knowledge of the "best" current evidence to discuss clinical management choices with patients in shared decision-making processes, incorporating patient priorities, values and preferences with resource utilization and other implementation factors. The "best" current evidence is characterized by high-quality clinical study designs that lead to unbiased results that can be generalized to specific patient-care situations. Integrating such evidence into shared decision-making discussions should lead to optimized patient-relevant outcomes.

References

1. Sackett DL, et al. Evidence based medicine: what it is and what it isn't. BMJ. 1996;312(7023):71–2.
2. Guyatt GH, et al. GRADE: an emerging consensus on rating quality of evidence and strength of recommendations. BMJ. 2008;336(7650):924–6.
3. Singh S, et al. Review article: a primer for clinical researchers in the emergency department: part X. Understanding economic evaluation alongside emergency medicine research. Emerg Med Australas. 2019;31(5):710–4.
4. Upadyhe S. Pitfalls in critical appraisal. In: Rowe B, editor. Evidence-based emergency medicine. Hoboken: Blackwell Books; 2008.
5. Oxford Centre for Evidence-Based Medicine. Centre for evidence based medicine. Univrsity of Oxford—resources. Available from https://www.cebm.ox.ac.uk/resources/resource
6. Baker GR, McGillion MH, Gavin F. Engaging with patients on research to inform better care. CMAJ. 2018;190(Suppl):S6–s8.
7. de Wit M, et al. Preparing researchers for patient and public involvement in scientific research: development of a hands-on learning approach through action research. Health Expect. 2018;21(4):752–63.
8. Frank L, et al. Engaging patients and other non-researchers in health research: defining research engagement. J Gen Intern Med. 2020;35(1):307–14.
9. Pelton L, Knihtila M. Experience-based co-design of health care services—implementation guide. Boston, MA: Institute for Healthcare Improvement; 2018.

Part II

Use of Cannabis in the Management of Specific Medical Conditions

The Effects and Benefits of Cannabis on the Gastrointestinal Disorders

8

Lawrence B. Cohen

Introduction

The endocannabinoid system plays an important role in the homeostasis and cellular function of the gastrointestinal (GI) tract [1]. Both cannabinoid receptors, CB_1 and CB_2 are present in the enteric nervous system and are specifically tied to cholinergic neurons. Additional receptors are also seen in the colonic epithelial cells and vascular smooth muscle cells of the colon. Given the vast distribution of these receptors in the GI tract and potential effects on activation, cannabinoids can have a multitude of effects including include nausea/vomiting, pain regulation, motility and regulation of inflammation. While several GI organizations have not fully approved the use of cannabis for any gastrointestinal, hepatologic or pancreatic diseases, new research is showing promise for specific conditions.

The GI Tract and Endocannabinoid System (See Chap. 3 on Pharmacology of Cannabinoids)

- The two main endocannabinoid receptors in the GI tract are CB_1 and CB_2 [1–3].
  - Type 1 (CB_1) receptors are located in the enteric nervous system (such as the epithelium of the GI tract) and sensory terminals of vagal and spinal neurons, regulating neurotransmitter release.

    CB_1 receptors are also seen in the smooth muscle cells of the colon.
  - Type 2 (CB_2) receptors are mostly distributed through the immune system producing a host of immunotherapeutic responses, including modifying inflammatory expression by macrophages, neutrophils, B and T cell subtypes.

L. B. Cohen (✉)
Sunnybrook Health Sciences Centre, Toronto, ON, Canada

University of Toronto, Toronto, ON, Canada
e-mail: Lawrence.Cohen@sunnybrook.ca

© Springer Nature Switzerland AG 2022
R. Valani (ed.), *Cannabis Use in Medicine*,
https://doi.org/10.1007/978-3-031-12722-9_8

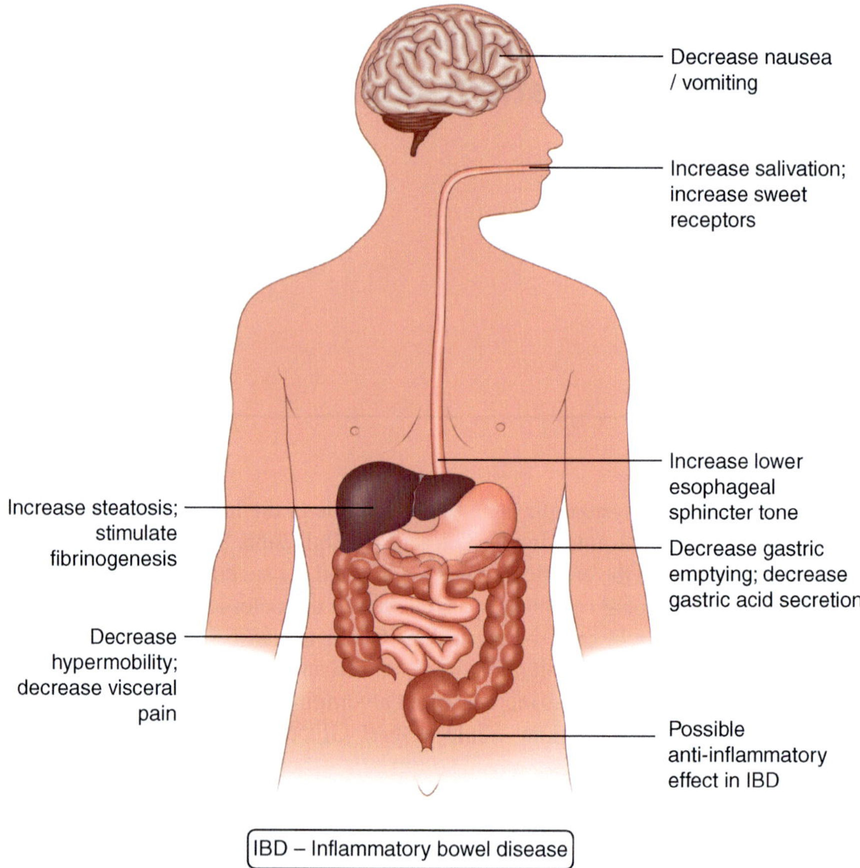

Fig. 8.1 Effects of cannabis on the gastrointestinal tract

- The GI tract can produce and metabolize its own ligands (Anandamide and 2-Arachidonoylglyerol) and as well as up- and down- regulate endocannabinoid receptors in order to facilitate appropriate bowel function [1].
- Effects on the GI tract include [3]: See Fig. 8.1.
    - Decrease in esophageal lower sphincter relaxation. This reduces emesis and provides antiemetic properties.
    - Decrease gastric acid secretion and gastric emptying.
    - Reduces hypermotility of the bowels associated with inflammatory diseases. This may potentially benefit diarrhea and other inflammatory aspects of inflammatory bowel disease.

- Decrease visceral pain. (See Chap. 11 on The use of Cannabis for pain management)
- Both CB_1 and CB_2 receptors inhibit GI muscle contraction through presynaptic decrease in excitatory neurotransmitter (acetylcholine and Substance P) release.

Effects of Cannabis on Specific GI Conditions

1: Anorexia and Weight Loss

- CB_1 receptors in the hypothalamus contribute to the regulation of appetite and energy balance.
- Studies on the efficacy of exogenous cannabinoids in modifying appetite and weight gain are controversial and may reflect differences in study design, the specific disease state being evaluated, and outcome parameters.
  - Strasser et al. reported no benefit from synthetic cannabinoids in malignant anorexia-cachexia compared to placebo controls; whereas, Brisbois et al. demonstrated significant improvement in appetite, enhancement of taste, and increased protein-calorie intake in cannabinoid treated cancer patients versus placebo treated control group [4, 5].
- Patients with AIDS associated anorexia and weight loss have had significant improvement in weight gain and quality of life with the use of cannabis [6].
- Cannabinoids may have also have a therapeutic role in weight reduction strategies.
  - Alshaarowy and Anthony recently showed an inverse relationship between cannabis use and obesity [7].
    The proposed mechanism of action is that chronic cannabis use may down regulate CB_1 receptors and upregulate CB_2 receptors in the hypothalamus leading to weight reduction.

2: Nausea/Vomiting

- Cannabis and related cannabinoids may be considered as primary or adjunctive therapy for limited periods in the management of refractory nausea and vomiting associated with chemotherapy, especially where conventional medications have been ineffective [8].
- CB_1 receptors are distributed throughout the brain, including the dorsal vagal complex of the brainstem (area postrema) which is involved in pathogenesis of vomiting [9].
  - A meta-analysis by Smith et al. concluded that cannabinoids yielded significant efficacy in the treatment of chemotherapy induced nausea and vomiting (CINV) [10].

- Despite evidence of its usefulness, the American and European oncology guidelines do not recommend cannabis be used as a first line agent in the management of CINV. Instead, they state that cannabinoids should be prescribed for CINV after conventional medical therapy has failed [11].
- The long-term effectiveness and safety of cannabis for chronic gastrointestinal symptoms, such as Irritable Bowel Syndrome and nausea/vomiting, remain unknown.
  - Until further studies of cannabis in the management of nonspecific GI symptoms should be avoided

3: Irritable Bowel Syndrome (IBS)

- The prevalence of IBS varies from 15 to 40% worldwide. Given the high prevalence and limited therapeutic options to manage the symptoms of IBS, cannabis offers an alternative therapeutic option.
- Cannabinoids reduce intestinal motility and secretions via CB_1 agonist activity [12].
- There are anecdotal and clinical studies reporting improved symptoms for a spectrum of GI issues such as IBS-constipation, diarrhea, anorexia, nausea, abdominal pain, providing impetus for use in motility disorders such as irritable bowel syndrome [13–16].
- The abdominal visceral pain of IBS is attributed to enhanced perception to colonic distention in about 70% of patients and that visceral sensation is mediated, in part, through the cannabinoid receptors [14].
  - Given the role of the cannabinoid receptor in IBS would suggest that cannabis may benefit these patients.
- The potential therapeutic role for cannabis in diarrhea- or pain-dominant forms of IBS is supported by small number of clinical trials.
  - Arguable efficacy was reported on abdominal pain perception, and changes in intestinal motility.
- A component of this CB_1 mediated effect on motility and prolonged gastric emptying has been shown to cause early satiety.
  - This has been suggested as a method of weight loss strategy, contrary to the well-known CNS mediated appetite stimulation and modifier of nausea.

4: Inflammatory Bowel Disease (IBD)

- The incidence of IBS varies by geography. It is estimated that 0.7% of Canadians and 1.3% of adults in the United States suffer from IBS.
  - The incidence is higher in Northern Europe and those with a family history.
- The pathophysiology is complicated, and both, a genetic predisposition as well as environmental factors contribute to the risk of having IBD.

- The pathophysiology of IBD is a complex interplay between:
  - Epithelial cells.
  - Immune cells—activation and uncontrolled inflammation within the bowel that is linked to unregulated T-cell activation.
  - Normal microbial flora—patients with IBD have a different composition of gut microbes which can modify the inherent epithelial cells.
- The two main forms of inflammatory bowel disease (IBD) are Crohn's Disease and Ulcerative Colitis. Both conditions have been linked to altered immune system responses.
  - Ulcerative Colitis (UC)

    UC is a chronic inflammatory process that extends from rectum proximally to cecum, and is limited to the colo-rectal mucosa in a confluent pattern.

    The inflammation is limited to the mucosal and superficial submucosal layers.

    The mucosa is granular, friable and ulcerated with edema, hemorrhage, with evolution of pseudo-polyps (inflammatory polyps) over time.
  - Crohn's Disease

    Crohn's Disease can affect any part of the gastrointestinal tract and usually spares the rectum.

    Crohn's Disease affects all the layers of the GI tract, with transmural inflammation that may be complicated by abscess and fistulae formation, featured by "cobble stoning" pattern of mucosal inflammation and deep serpiginous ulceration.

    Focal crypts and abscesses are seen on microscopy.
- Non-GI manifestations of IBD include:
  - Dermatologic—erythema nodosum and pyoderma gangrenosum.
  - Musculoskeletal—arthritis, ankylosing spondylitis, sacroiliitis.
  - Ophthalmic—uveitis, iritis, episcleritis.
  - Hepatobiliary—primary sclerosing cholangitis (PSC)
- The gastrointestinal tract has an extensive network of CB_2 receptors that may promote the integrity of intestinal epithelium. They potentially mediate significant mucosal anti-inflammatory effects, which are known to be predisposing factors in IBD [17, 18].
- Cannabidiol (CBD) has been shown to improve both inflammatory signs and symptoms of IBD.
  - CBD purportedly exerts an anti-inflammatory effect by stimulating the peroxisome proliferator activated receptor gamma (PPAR-gamma) [19].

    A prospective cohort study reported that the majority of their IBD patients found cannabis to be very helpful in completely relieving abdominal pain, nausea and diarrhea taken in conjunction with their prescription anti-inflammatory medications [20].

    Objective parameters of clinical, laboratory, and endoscopic improvement in IBD (Crohn's Disease) patients have been described and demonstrated [21, 22].

Interestingly, there may be a difference in response between patients with UC and Crohn's Disease.
- There may be worse outcome in Crohn's Disease patients receiving CBD compared to the UC group [23].
- Cannabis therapy may be considered an alternative adjunct to conventional therapeutics in the management of the signs and symptoms of active IBD, but not as primary anti-inflammatory therapy [24]
- To date there is insufficient evidence for cannabis to be effective to alter the course of disease in patients with inflammatory bowel disease. Further studies are needed to provide evidence on the role of cannabis on the natural history of IBD [25]

5: Liver Diseases

- There is a growing body of evidence to suggest that cannabinoids influence a variety of liver disorders, including hepatic steatosis and fibrosis, portosystemic encephalopathy, and alcoholic liver disease [24].
- Stimulation of CB_1 receptors in the liver may promote steatosis via increasing lipogenesis, decreasing fatty acid oxidation and inducing hyperphagia. On the other hand, CB_1 antagonists suppresses hepatic steatosis [25].
  - Hepatic CB_1 receptors may also stimulate fibrogenesis especially in alcohol hepatitis, and in-vitro and in-vitro studies showed that CB_1 antagonists may protect against development of alcohol induced liver fibrosis [26].
- The effects of daily cannabis use in viral hepatitis patients is controversial.
  - Ishida et al. found daily cannabis use to be strongly associated with moderate to severe fibrosis in hepatitis C patients [27].
  - Brunet et al. and Liu et al. discovered no adverse effects of cannabis use on the natural history of Hepatitis C Virus (HCV) [28].
  - There are no reported data on the impact of cannabis on the natural history of Hepatitis B infection.
  - In patients co-infected with HIV and Hepatitis C, cannabis use may reduce the rate of steatosis as well as insulin resistance. The impact on fibrogenesis in the co-infected group is controversial [29, 30].
- Stimulation of CB_2 receptors, which may be upregulated in chronic liver disease, have been reported to protect against hepatic fibrosis.
- Interestingly, in alcoholic liver disease, the balance of cannabinoids may have a protective effect by reducing oxidative stress that leads to inflammation and steatosis thereby resulting in lower rates of alcohol steato-hepatitis, cirrhosis and hepatocellular carcinoma.
- Epidemiological studies suggest that cannabis use is associated with a lower prevalence of non-alcohol fatty liver disease.
- While cannabis hepatotoxicity is arguable, cannabinoids may have a defined role in the management of chronic liver disease as more studies emerge.

6: Pancreatitis

- The pancreas contains both CB_1 and CB_2 receptors.
  - As in liver, activation of CB_1 pancreatic receptors promotes fibrogenesis, opposed by CB_2 receptor agonists.
- Cannabis induced pancreatitis has been reported and may be dose and duration of use dependent.
  - Treatment is the same as other causes of medical pancreatitis.

7: Cannabinoid Hyperemesis Syndrome (CHS) (See Chap. 12 for More Details on Management and Treatment)

- There is an increased incidence of CHS seen in the emergency department. The prevalence is difficult to determine given the variability in symptoms and use of cannabis, as well as underreporting of this condition.
- CHS is most commonly seen in patients regularly use cannabis. It presents with protracted nausea and vomiting.
- While low dose CBD yields anti-emetic properties, higher doses have a pro-emetic effect. Hence the reason why it is seen commonly in chronic users.
- The clinical course follows three phases:
  - ***Prodromal phase***: This phase presents with early morning nausea, fear of vomiting and non-specific abdominal discomfort. The prodromal phase can last for months or years.
  - ***Hyperemetic phase***: This phase begins with the development of intense nausea, pernicious vomiting and diffuse abdominal pain. This is when patients present to the Emergency Department as they cannot control their vomiting with typical anti-emetic agents.

    Patients may propagate this phase by continuing to consume cannabis for the misbelief that they need the antiemetic property of the drug.

    See section in Chap. 12 on specific treatment for CHS in the Emergency Department.
  - ***Recovery phase***: This phase is highlighted by improving symptoms and signs described above, weeks to months after withdrawing from cannabis consumption, with progressive weight regain as a result of a return to normal mood and eating patterns.

Conclusion

- The principles of pharmacotherapy, understanding the risk:benefit ratio, and the efficacy of a compound towards a disease applies to cannabis as it does to any drug. Therefore, rigorous clinical trials need to be designed and undertaken to answer the clinical questions of appropriate indications for medical cannabis,

therapeutic dosing and appropriate monitoring for effectiveness and adverse effects.
- These principles are being applied for promising uses of cannabis in gastrointestinal, hepatic, and pancreatic disorders.
  - As more well designed clinical research trials are being conducted, rational prescribing profiles will be available.
- The Canadian Association of Gastroenterology recommends "that physicians in Canada familiarize themselves with important aspects of medical cannabis before authorizing a patient for medical use. Moreover, with recreational use being so common, it also behooves physicians to understand the risks involved for patients and to be able to counsel them accordingly" [31].
- A comprehensive summary of the role of cannabis in gastrointestinal, hepatic and pancreatic diseases, and effects on metabolic disorders such as obesity, have been published by Gotfried et al. [32].

References

1. Gerich ME, et al. Medical marijuana for digestive disorders: high time to prescribe? Am J Gastroenterol. 2015;110(2):208–14.
2. Massa F, Monory K. Endocannabinoids and the gastrointestinal tract. J Endocrinol Investig. 2006;29(3 Suppl):47–57.
3. Uranga JA, Vera G, Abalo R. Cannabinoid pharmacology and therapy in gut disorders. Biochem Pharmacol. 2018;157:134–47.
4. Strasser F, et al. Comparison of orally administered cannabis extract and delta-9-tetrahydrocannabinol in treating patients with cancer-related anorexia-cachexia syndrome: a multicenter, phase III, randomized, double-blind, placebo-controlled clinical trial from the Cannabis-In-Cachexia-Study-Group. J Clin Oncol. 2006;24(21):3394–400.
5. Brisbois TD, et al. Delta-9-tetrahydrocannabinol may palliate altered chemosensory perception in cancer patients: results of a randomized, double-blind, placebo-controlled pilot trial. Ann Oncol. 2011;22(9):2086–93.
6. Haney M, et al. Dronabinol and marijuana in HIV-positive marijuana smokers. Caloric intake, mood, and sleep. J Acquir Immune Defic Syndr. 2007;45(5):545–54.
7. Alshaarawy O, Anthony JC. Are cannabis users less likely to gain weight? Results from a national 3-year prospective study. Int J Epidemiol. 2019;48(5):1695–700.
8. Whiting PF, et al. Cannabinoids for medical use: a systematic review and meta-analysis. JAMA. 2015;313(24):2456–73.
9. Sharkey KA, Darmani NA, Parker LA. Regulation of nausea and vomiting by cannabinoids and the endocannabinoid system. Eur J Pharmacol. 2014;722:134–46.
10. Smith LA, et al. Cannabinoids for nausea and vomiting in adults with cancer receiving chemotherapy. Cochrane Database Syst Rev. 2015;2015(11):Cd009464.
11. Tafelski S, Häuser W, Schäfer M. Efficacy, tolerability, and safety of cannabinoids for chemotherapy-induced nausea and vomiting—a systematic review of systematic reviews. Schmerz. 2016;30(1):14–24.
12. Aviello G, Romano B, Izzo AA. Cannabinoids and gastrointestinal motility: animal and human studies. Eur Rev Med Pharmacol Sci. 2008;12(Suppl 1):81–93.
13. Adejumo AC, et al. Relationship between recreational marijuana use and bowel function in a nationwide cohort study. Am J Gastroenterol. 2019;114(12):1894–903.

14. Malik Z, Baik D, Schey R. The role of cannabinoids in regulation of nausea and vomiting, and visceral pain. Curr Gastroenterol Rep. 2015;17(2):429.
15. Wong BS, et al. Pharmacogenetic trial of a cannabinoid agonist shows reduced fasting colonic motility in patients with nonconstipated irritable bowel syndrome. Gastroenterology. 2011;141(5):1638–47.e1–7.
16. Wong BS, et al. Randomized pharmacodynamic and pharmacogenetic trial of dronabinol effects on colon transit in irritable bowel syndrome-diarrhea. Neurogastroenterol Motil. 2012;24(4):358–e169.
17. Hasenoehrl C, et al. The gastrointestinal tract—a central organ of cannabinoid signaling in health and disease. Neurogastroenterol Motil. 2016;28(12):1765–80.
18. Couch DG, et al. Cannabidiol and palmitoylethanolamide are anti-inflammatory in the acutely inflamed human colon. Clin Sci (Lond). 2017;131(21):2611–26.
19. Esposito G, et al. Cannabidiol in inflammatory bowel diseases: a brief overview. Phytother Res. 2013;27(5):633–6.
20. Ravikoff Allegretti J, et al. Marijuana use patterns among patients with inflammatory bowel disease. Inflamm Bowel Dis. 2013;19(13):2809–14.
21. Naftali T, et al. Treatment of Crohn's disease with cannabis: an observational study. Isr Med Assoc J. 2011;13(8):455–8.
22. Naftali T, et al. Cannabis induces a clinical response in patients with Crohn's disease: a prospective placebo-controlled study. Clin Gastroenterol Hepatol. 2013;11(10):1276–1280.e1.
23. Storr M, et al. Cannabis use provides symptom relief in patients with inflammatory bowel disease but is associated with worse disease prognosis in patients with Crohn's disease. Inflamm Bowel Dis. 2014;20(3):472–80.
24. Goyal H, et al. Cannabis in liver disorders: a friend or a foe? Eur J Gastroenterol Hepatol. 2018;30(11):1283–90.
25. Goyal H, et al. Role of cannabis in digestive disorders. Eur J Gastroenterol Hepatol. 2017;29(2):135–43.
26. Patsenker E, et al. Cannabinoid receptor type I modulates alcohol-induced liver fibrosis. Mol Med. 2011;17(11–12):1285–94.
27. Ishida JH, et al. Influence of cannabis use on severity of hepatitis C disease. Clin Gastroenterol Hepatol. 2008;6(1):69–75.
28. Brunet L, et al. Marijuana smoking does not accelerate progression of liver disease in HIV-hepatitis C coinfection: a longitudinal cohort analysis. Clin Infect Dis. 2013;57(5):663–70.
29. Nordmann S, et al. Daily cannabis and reduced risk of steatosis in human immunodeficiency virus and hepatitis C virus-co-infected patients (ANRS CO13-HEPAVIH). J Viral Hepat. 2018;25(2):171–9.
30. Carrieri MP, et al. Cannabis use and reduced risk of insulin resistance in HIV-HCV infected patients: a longitudinal analysis (ANRS CO13 HEPAVIH). Clin Infect Dis. 2015;61(1):40–8.
31. Andrews CN, et al. Canadian Association of Gastroenterology Position Statement: use of cannabis in gastroenterological and hepatic disorders. J Can Assoc Gastroenterol. 2019;2(1):37–43.
32. Gotfried J, Naftali T, Schey R. Role of cannabis and its derivatives in gastrointestinal and hepatic disease. Gastroenterol. 2020;159:62–80.

Neurological Diseases and Cannabinoid Treatment

9

Magda Nowicki, Rahim Dhalla, Richard Huntsman, Jane Alcorn, and Evan Cole Lewis

Introduction

There are many neurological conditions for which there is no cure, and patients are offered treatment options to reduce symptoms and improve quality of life. With a better understanding of the endocannabinoid system, there is the potential to use cannabis as adjunct medication or as an alternative to current treatment recommendations. While cannabis has been evaluated in select neurologic disorders, it has the potential to be used in other conditions as well.

This chapter reviews three neurological conditions that have shown some promise with cannabis: epilepsy, multiple sclerosis, and sleep disorders. With ongoing research, the use of cannabis for these and other conditions can be explored.

M. Nowicki
The Hospital for Sick Children, University of Toronto, Toronto, ON, Canada

University of Toronto, Toronto, ON, Canada
e-mail: magda.nowicki@sickkids.ca

R. Dhalla
Hybrid Pharm, Ottawa, ON, Canada
e-mail: rahim@hybridpharm.com

R. Huntsman
Royal University Hospital, University of Saskatchewan, Saskatoon, SK, Canada
e-mail: dr.huntsman@usask.ca

J. Alcorn
College of Pharmacy and Nutrition, University of Saskatchewan, Saskatoon, SK, Canada
e-mail: jane.alcorn@usask.ca

E. C. Lewis (✉)
Neurology Centre of Toronto, Toronto, ON, Canada

University of Toronto, Toronto, ON, Canada
e-mail: Evan.Lewis@neurologycentretoronto.com

© Springer Nature Switzerland AG 2022
R. Valani (ed.), *Cannabis Use in Medicine*,
https://doi.org/10.1007/978-3-031-12722-9_9

Background: Cannabis Formulations and Products

- Because the nomenclature describing cannabis products is not standardized, it makes it difficult to undertake adequate reviews or compare different clinical studies. See Chap. 7 on Evidence based reviews on cannabis therapy.
- For the purpose of understanding the dosing regimen and interpreting outcomes described in this section, the following terms are utilized:
  - Synthetic cannabinoid: these are compounds that are synthesized in the lab.
    They usually consist of a single cannabinoid of choice such as Δ9-Tetrahydrocannabinol (THC).
    Examples of synthetic cannabinoids include Dronabinol (Merinol—withdrawn in Canada) and Nabilone (Cesamet).
  - Purified cannabinoids: these are mixtures that are extracted from the cannabis plant.
    Also referred to as "cannabinoid isolate" or pharmaceutical grade compounds. For example, 99% Cannabidiol (CBD) isolate can be extracted from the cannabis plant.
    The extract usually contains minimal amounts of other cannabinoids (<1%). A good example of this is Epidiolex which is a CBD oral solution approved by the FDA and the EMA for the treatment of Lennox-Gastaut syndrome and Dravet syndrome for patients over the age of 2 years.
  - Cannabinoid-Rich extract: these are also mixtures extracted from the cannabis plant.
    Also referred to as "cannabinoid extract" or "artisanal" products
    These extracts contain a predominance of one or more cannabinoids of choice (e.g. CBD-rich, balanced THC:CBD, etc.)
    The most common ratios may include 1:20 THC:CBD, 20:<1 THC:CBD, and 10:10 THC:CBD.
    - Nabiximols (2.7:2.5 THC: CBD) -sold as an oral spray named Sativex- is an example of a very specific extract mixture that has been used to treat MS
    Technically, any ratio of any cannabinoid can be developed by extracting a sample and adding an appropriate isolate.

Neuropharmacology (See Chap. 2 on the Pharmacology of Cannabis)

- The neuropharmacology of the endocannabinoid system (ECS) and its interaction with phytocannabinoids is complex and extensive.
- Cannabinoid receptors (especially CB_1) are ubiquitous in the mammalian brain, accounting for many of the reported neurological effects of cannabis.
- The distribution of CB_1 receptors in the central nervous system (CNS) has been well classified [1, 2]:

- There is a high concentration in the hippocampal complex, entorhinal and cingulate cortices, frontal gyrus, amygdaloid complex, substantia nigra and the cerebellar molecular layer.
- Lower densities are found in the thalamic and hypothalamic nuclei, brainstem (except dorsal motor nucleus of the Vagus) and spinal cord.
- CB_1 receptors can be found at both, pre- and post- synaptic neurons, as well as excitatory and inhibitory neurons.
- CB_2 receptors predominate in cardiovascular system, GI tract, liver, adipose tissue, bone, and reproductive system [3, 4].
  - There have been CB2 receptors identified in the brain, however, their exact role is uncertain.
  - There is some evidence to support that CB2 receptors may play a role in neuronal signaling.
- The role of the ECS in the CNS is an area of active research. The postulated function of the ECS in the CNS relates to neuronal transmission [3, 5, 6].
  - The ECS accomplishes this by retrograde signaling. See Fig. 9.1.
- Endocannabinoids, Anandamide (AEA) and 2-Arachidonoylglycerol (2-AG), are produced from membrane lipids in the post-synaptic neuron in response to an action potential.
- AEA and 2-AG diffuse in retrograde fashion to the pre-synaptic terminal and act on presynaptic CB_1 (G-protein coupled receptor) which results in decreased vesicular neurotransmitter release into the synapse by:
  - 1) Inhibition of adenylyl cyclase: decreases formation of cAMP and protein kinase A and inhibition of Ca^{2+} transport into pre-synaptic terminal.
  - 2) Opening of inwardly-rectifying K^+ channels: causes hyperpolarization of pre-synaptic neuron.
- AEA and 2-AG are degraded in the pre- and post-synaptic terminals.
  - AEA degradation: fatty acid amide hydrolase (FAAH) breaks down AEA into arachidonic acid (AA) and ethanolamine (ETA).
  - 2-AG degradation: monoacylglycerol lipase (MAGL) breaks down 2-AG into AA and glycerol.
- The final common pathway of AEA and 2-AG agonism at the pre-synaptic CB_1 receptor is:
  - Reduction in neurotransmitter released into the synaptic cleft and, thus, *decrease in action potential transmission.*
  - Because this effect can occur with *excitatory* and *inhibitory* neurons, the ECS works to regulate neuronal signaling via the action of AEA and 2-AG at the CB_1 receptor.
- Exogenous cannabinoids such as Δ9-tetrahydrocannabinol (THC) and synthetic CB_1 agonists like nabilone or dronabinol act at the CB_1 receptor and have a similar effect with varying magnitude on neurotransmitter release as described above.
- The downstream effect is identical to the endocannabinoids but the effect is widespread as these exogenous compounds (e.g. THC) are not specific to particular neurons or locations within the brain.

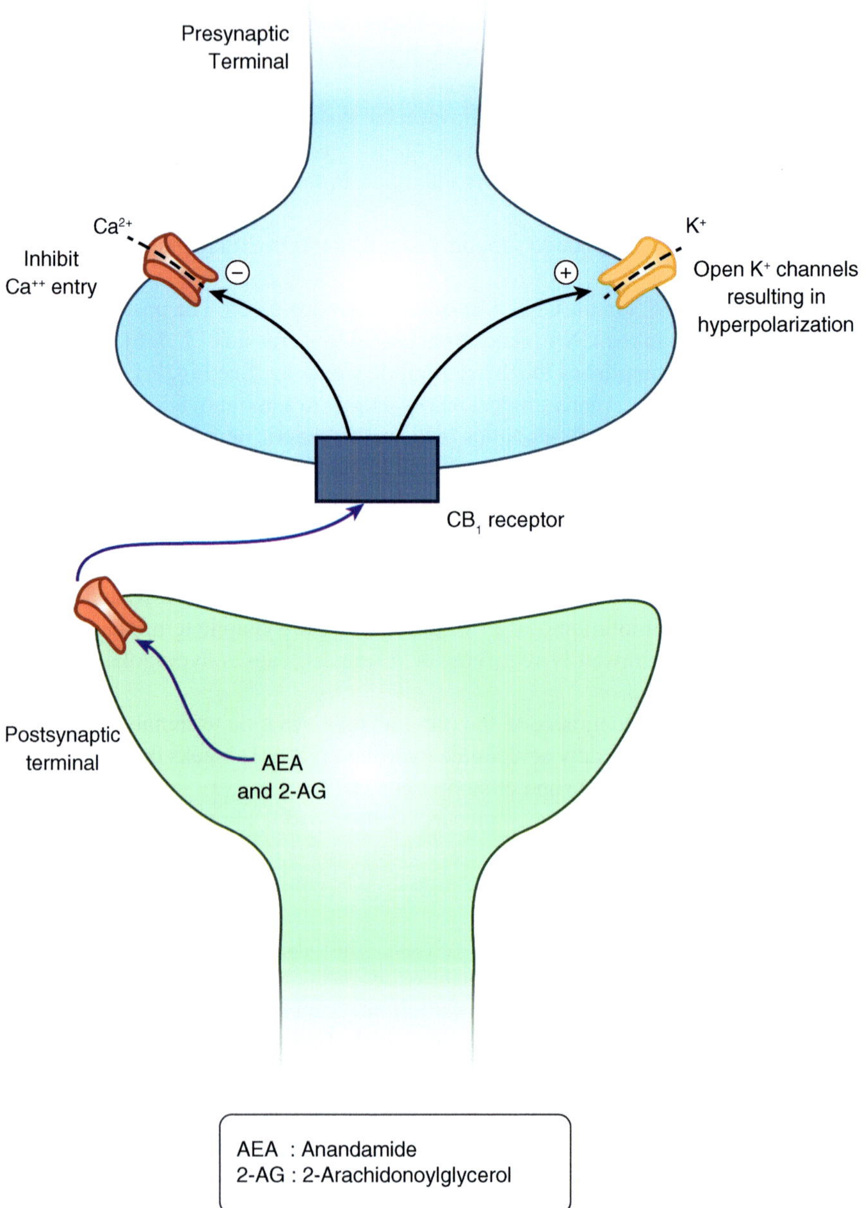

Fig. 9.1 The endocannabinoid system

Epilepsy

- The ECS regulates neuronal excitability, and modulation of the ECS is hypothesized as the basis for treatment in epilepsy.
- There is high quality evidence for medical cannabis as a therapeutic agent available for pediatric epilepsy. Cannabis has been shown to decrease the frequency of seizures compared to placebo [7].

The Effects of Seizure Control with Cannabis

- Purified CBD and its role in management of epilepsy:
    - The only approved purified CBD formulation in North America and Europe is Epidiolex, which is a form of purified CBD (~99%) with minimal <1% THC, minor cannabinoids and non-cannabinoid components.
        In the US, it is approved for tuberous sclerosis (TSC), Lennox Gastaut syndrome (LGS) and Dravet syndrome. It has been approved in Europe for LGS and Dravet (but only with concomitant use of clobazam).
    - A double-blind placebo-controlled trial by Miller et al. evaluated efficacy and safety of CBD at 10 mg/kg/day in Dravet syndrome compared to 20 mg/kg/day for 14 weeks [8].
        Both doses showed similar effectiveness in seizure reduction.
    - In patients with Dravet syndrome, 20 mg/kg/day of CBD significantly reduced monthly frequency of seizures (from 12.4 to 5.9) compared to controls [9].
    - In patients with LGS treated with 20 mg/kg/day of CBD vs. placebo, the monthly frequency of atonic seizures reduced by 43.9% in the CBD group [10].
        Lower doses (as low as 10mg/kg/day) also appeared to be effective in another RCT [11].
    - Purified CBD is also effective in other epilepsy types.
        In addition to LGS and Dravet syndrome, open label studies evaluated purified CBD in patients with various types of genetic epilepsies and genetic disorders associated with seizures such as Febrile infection-related epilepsy syndrome (FIRES), TSC, Doose syndrome, CDKL5 deficiency and Aicardi syndrome.
    - There has been expanding interest in the efficacy of cannabis in other severe, drug resistant epileptic encephalopathies (e.g. infantile spasms). Further data is still pending to make appropriate recommendations.
    - The effect of purified CBD in focal epilepsy is unclear.
        CBD in patients with refractory temporal lobe epilepsy (TLE) first studied by Cunha et al. resulted in improvement of seizures [12].
        Pooled data from open label studies also reports focal seizure reduction with use of cannabis-based products (30–90% reduction) at 8 weeks and up to >16 months treatment duration [7].

- CBD-rich Extracts and Epilepsy
    - There is limited data regarding dosing and safety of CBD-rich extracts in the management of seizure disorders.
    - A retrospective study looked at refractory epilepsy treated with CBD-rich extracts (1:20 THC:CBD) for an average of 6 months. CBD treatment significantly decreased seizures (89% reported seizure frequency reduction, 43% had >50% reduction) [13].
        CBD dosing was 1–20 mg/kg/day, with the majority being <10 mg/kg/day
    - A prospective open label study by McCoy et al. examined the effect of a CBD-rich extract (2:100 THC:CBD) in 20 children with Dravet. CBD dosing ranged from 2 to 16 mg/kg/day, mean dose was 13.3 mg/kg/day [14].
        There was a significant (70.6% median) monthly reduction in motor seizures.
    - Meta-analysis of observational clinical studies of cannabidiol-based products in children with refractory epilepsy showed that more patients taking CBD-rich extracts reported improvement in seizure control versus those taking purified CBD (71 vs. 46% with statistical significance, p < 0.0001). However, there was no difference in those reaching >50% seizure reduction. There were fewer side effects in patients taking CBD-rich extracts compared to those taking purified CBD (p < 0.0001).

Other Considerations

- There is a paucity of long-term data on the efficacy of CBD which may add to the reluctance for its use.
- Cannabis does interact with other anti-epileptic medications:
    - CBD inhibits the metabolism of clobazam and its metabolite n-desmethylclobazam resulting in increased levels.
    - There has been one retrospective study that examined the efficacy of CBD-rich extracts with and without clobazam.
        There was no difference in seizure reduction in patients on CBD that were not on clobazam, suggesting the response to CBD was independent of clobazam use [15].

Adverse Effects

- The reported adverse effects of exposure to high amounts of CBD (purified CBD and CBD-rich extracts) are similar in RCTs and open label trials.
- The most common clinical adverse events (defined as those seen in greater than 10%) are somnolence, diarrhea, decreased appetite and fatigue.
    - Most of these effects are mild to moderate in severity.

- Serious side effects are rare (1–7%) and these include respiratory tract infections, weight loss, and seizure exacerbation.
- Adverse effects tend to decrease and stabilize after 12 weeks.
- The most common biochemical adverse effect is elevation of liver enzymes.
  - Elevated liver enzymes are associated with concurrent valproate use, high dose CBD (>20 mg/kg/day) or those who had elevated transaminases before starting CBD.
  - Most elevations are transient and resolve with cessation of medical cannabis.

Summary Points

- RCTs demonstrated that purified CBD reduces seizure frequency in specific epilepsies, including Dravet syndrome, LGS and TSC.
  - There is insufficient data to date on the use of CBD in other types of epilepsy.
- Observational trials utilizing CBD-rich extracts have also showed positive results in treatment of epilepsy and may have a better side effect profile than purified CBD.
- The most common adverse events with CBD include: diarrhea, fatigue and appetite suppression.
  - These events are mild-moderate in severity and are more frequent with high doses of CBD.

Multiple Sclerosis

- Multiple sclerosis (MS) is a chronic demyelinating disease of the CNS.
  - MS is characterized by demyelinating plaques in grey and white matter that leads to progressive neuroaxonal loss.
- The most distressing and difficult to treat symptoms in multiple sclerosis include spasticity, painful spasms and neuropathic pain.
- Reports of patients experiencing symptomatic relief after smoking cannabis prompted some of the earlier research in the medical cannabis field.
- Most studies have looked at the efficacy of Nabiximols and other oral cannabis-rich extracts to treat pain and spasticity.
  - Nabiximols (an oral-mucosal cannabis rich-extract spray that contains 2.7 mg THC and 2.5 mg CBD per spray) was developed to treat MS symptoms.

Efficacy of Cannabis for Symptom Relief in Patients with MS

- Pain and Spasms. (See Chap. 11 on The use of cannabis for pain management)
  - A study by Rog et al. showed decreased pain intensity in patients taking nabiximols [16].

- The CAMS trial was a randomized placebo-controlled trial of 667 patients for treatment of pain and spasms [17].

  Oral cannabis-rich extract Cannador (2.5 mg/ml THC and 1.25 mg/ml CBD) vs. synthetic THC preparation for 15 weeks. The maximum dose was 25 mg of THC.

  - Pain and muscle spasm improvement after 14 weeks of treatment observed in both cannabis-rich extract and synthetic THC groups.

- Other lower quality studies have produced mixed results on the efficacy of nabiximols in central pain. Many of the studies had limited power, strong placebo effect or were of lower quality.

- Spasticity
  - Whiting et al. reviewed studies that looked at the effect of various formulations of cannabis on spasticity [18].

    They concluded that there is moderate quality evidence suggesting cannabinoids were associated with improvements in spasticity.

    Although there was no significant improvement on objective measurements of spasticity (using the Ashworth scale) compared to placebo, patients reported subjective improvements in spasticity when using nabilone and nabiximols.

  - Koppel et al. also concluded there is evidence supporting the use of cannabis in spasticity [19].
  - The CAMS trial showed no effect on Ashworth scale scores but there was a treatment effect on patient-reported spasticity and pain [17].
  - Overall, cannabis does not appear to show improvements on objective measures of spasticity, however there are significant improvements of patient-reported spasticity.

Other Outcomes and Adverse Effects

- There has been no strong evidence supporting cannabis use in modifying disease progression, reducing ataxia, tremor, or improving bladder dysfunction in patients with MS.
- The most common reported adverse effects are mild-moderate severity, including dizziness, diarrhea, nausea, fatigue, cognitive disturbances and headache.
  - These adverse effects are likely dependent on each individual patient's underlying condition (e.g. cognitive effects may be more prominent in those with underlying cognitive impairment) [18–20].

Summary Points

- Cannabinoids (nabiximols and other cannabis-rich extracts) may be beneficial in reducing spasticity and pain in MS.
- Patients have reported benefits with reduced spasticity and improvements in pain.
- The research is inconclusive with regards to treating other common symptoms such as ataxia, tremor and bladder dysfunction.

Sleep

- Normal sleep consists of rapid eye movement (REM) sleep and non-REM (NREM) sleep, and over a period of sleep they alternate cyclically.
  - NREM usually consists of the majority of sleep time for most individuals. It is further divided into three stages: N1 → N2 → N3/N4 (slow wave sleep).
  - REM, N2 and slow wave sleep are associated with restorative properties.
- The sleep-wake cycle is highly dependent on individual factors and influenced by chronic pain, mental health, stress and other comorbid medical issues.
- ECS plays a role in regulating sleep-wake balance and is implicated in the amount of time spent in each stage of sleep

Sleep Outcomes

- Cannabis extracts have been used for their sedative and hypnotic properties since the early 1900s [21].
- The early studies related to cannabis and sleep are largely observational. These studies evaluated sleep in chronic daily, recreational users of smoked cannabis.
  - Generally, these studies are of poor quality in that they lack control for confounding factors, use variable doses of THC, while the CBD and terpene profiles are not always reported.
  - Not surprisingly, the results are inconsistent on subjective and objective measures of sleep quality (e.g., sleep stages).
  - Across these studies, a common feature was a reduction in the sleep latency period.
- Synthetic forms of THC (such as nabilone and dronabinol) were assessed in patients with chronic medical issues, pain and anxiety.
  - There were subjective reports of improved sleep due to reduced nighttime awakenings consistently reported.
  - It is unknown whether the findings were secondary to cannabis effects on sleep or alleviation of anxiety and pain.
- An early study by Carlini et al. prompted interest in the use of CBD in sleep [21].
  - Patients were treated with gelatin capsules of either (1) 5 mg nitrazepam (2) CBD (40 mg, 80 mg, or 160 mg) or (3) placebo.
  - Patients reported increased sleep quality (due to increased sleep duration) with 160 mg CBD prior to sleep. All three doses of CBD decreased dream recall, suggesting fewer nighttime awakenings.
  - No adverse effects were reported.
  - Since the 2000s, more studies have begun examining the effects of controlled quantities of the most common cannabinoids (THC and CBD) on sleep.

Effect on Measures of Sleep

- Some evidence suggests that THC and CBD may have beneficial balancing effects on sleep.
- Nicholson et al. looked at healthy young adults between the ages of 18–35 [22].

- Subjects were divided into four treatment groups: (1) 15 mg THC (2) balanced THC and CBD (5 mg, 5 mg) or (3) higher dose balanced THC and CBD (15 mg, 15 mg) and (4) placebo.

  15 mg THC group: subjective reports of morning sleepiness, reduced latency to sleep in the early morning (objective measure of sleepiness), and memory impairment on objective testing.

  15 mg/15 mg CBD/THC group: no decreased latency to sleep in the early morning (reduced sleepiness) and no memory impairment.
  - THC may have sedative properties and CBD may have counteracting alerting properties at these doses.
  - Equal doses of CBD may counteract the residual daytime adverse effects of THC.

Effect on Sleep Architecture

- Studies on the effect of THC on sleep architecture have revealed mixed results.
  - Some suggest a differential effect of THC based on the dose.

    Low doses: increases slow wave sleep.

    High doses: decrease REM and slow wave sleep.
- The mode of delivery of cannabis also has a differential impact for individuals suffering from sleep difficulties.
  - For example, Vigil et al. compared ratings of self-perceived sleep difficulties in 409 individuals with previous reported insomnia [23].

    The use of pipes and vaporizers was associated with improved sleep and fewer adverse effects compared to smoking joints.

REM Sleep Disorder

- This is a disorder characterized by the loss of muscle tone during REM sleep, nightmares and active behaviour during dreaming.
- One study to-date has evaluated the effect of CBD in REM sleep behaviour disorder (RBD) and showed improvement in symptoms [24].
  - The proposed mechanism is that CBD may act on CB_1 receptors expressed on cholinergic-expressing neurons in sleep related brain areas. This leads to acetylcholine release-mediated improvement in RBD symptoms.

Summary Points

- It is likely that the ECS is involved in the regulation of the sleep-wake cycle.
- Sleep effects appear to depend on type of cannabinoid, dose and route of administration.

- THC potentially has short term benefits at lower doses including decreased sleep latency and increased slow wave sleep.
- CBD has differential effects based on dose:
  - Low dose CBD may have stimulating effects that counteract adverse effects of THC when used in combination.
  - Medium to high dose CBD appears to have a sedating effect, minimal adverse effects and no sleep architecture disturbances.
- CBD has demonstrated sustained improvements over weeks in REM behaviour disorder symptoms.

Important Cannabis Drug-Drug Interactions in Neurology

- Drug interactions can be summarized based on two principles:
  - pharmacodynamic (what the drug does to the body) and
  - pharmacokinetic (what the body does to the drug).
- The following tables summarize cannabinoid metabolism (Table 9.1), relevant known cannabis-drug interactions (Table 9.2), and disease specific interactions (Table 9.3).

Table 9.1 Metabolism of cannabinoids through the cytochrome p450 pathway

Cannabinoid	Substrate	Inhibits	Induces
CBD	1A2,2C9,2C19*,2D6*,3A4*	1A2,2C19, 3A4,	1A2 (inhalation)
THC	2C9*, 2C19*, 3A4*	1A2, 2C9, 2C19, 3A4	1A2 (Inhalation)

Table 9.2 Common drug interactions with cannabis

Drug	Metabolized by	Summary points	Management
Clobazam	3A4, 2C19	CBD increases plasma levels of clobazam, via 3A4 and 2C19 inhibition, and its metabolite, n-desmethylclobazam by 60% and 500%, respectively	Dose reduction of clobazam may be required
Phenytoin	2C9, 2C19, 3A4	CBD and THC inhibit 2C9, 2C19 and 3A4. May enhance anticonvulsant effects of phenytoin or possible risk of toxicity	Monitor blood levels of phenytoin and adjust dosing as necessary
Valproic Acid and Derivatives	2C9, 2C19	Cannabinoids may enhance hepatotoxic effects of valproic acid	Monitor ALT, AST and adjust dosing accordingly
Carbamazepine	3A4	May decrease serum concentrations of CBD and THC.	Adjust cannabinoid dosing
Warfarin	2C9	THC may inhibit metabolism and thus increase INR	Monitor INR. Caution patients on the signs and symptoms of bleeding

                                                                                                M. Nowicki et al.

Table 9.3 Linking disease specific cautions for cannabis use

Disease state	Drug	Metabolism	Summary points
Autism Spectrum Disorder, Attention Deficit Hyperactivity Disorder, Tourette's syndrome	Fluoxetine (Es)Citalopram	2C19 inhibitor 2C19, 3A4 Substrate 2D6 Substrate	May increase serum concentrations of cannabinoids Cannabinoids may increase serum concentration of citalopram. Dose reduction may be required Monitor for increased adverse effects of Risperidone Smoking may reduce serum concentrations
	Risperidone Olanzapine Stimulants	1A2, 2D6 Substrate Pharmacodynamic	Monitor for increased tachycardia
Epilepsy	Brivaracetam (Brivlera) Tiagabine Lacosamide Lamotrigine Levetiracetam Topiramate Eslicarbazepine Vigabatrin	2C19 Substrate 3A4 Substrate Pharmacodynamic	Cannabinoids may inhibit 2C19 and 3A4, increasing the effects of Brivaracetam & Tiagabine Monitor for adverse effects Monitor for increased CNS depressant effects
Multiple Sclerosis	Siponimod Teriflunomide	2C9 substrate 1A2 Inducer	THC may inhibit 2C9. Monitor for increased adverse effects of Siponiimod Smoking cannabis may reduce serum concentrations
Insomnia	Zopiclone Trazodone Quetiapine Diphenhydramine	3A4 3A4, 2D6 3A4 2C9, 2C19	Cannabinoids may inhibit 3A4. Increase pharmacodynamic effects and anticholinergic effects. Dose reduction may be required

Summary

The use of cannabis in neurological disorders is an active area of research. Research is showing that the ECS has a role in modulation of neuronal signaling, with implications in the pathogenesis and treatment of CNS disease. In patients with epilepsy, purified CBD reduces seizure frequency in specific epilepsies, and there have been positive results in the use of CBD-rich extracts with fewer side effects compared to traditional antieplietic agents. Cannabinoids may also be beneficial in reducing spasticity and pain in patients with MS. In addition, cannabis may have a role in improving sleep. However, the effects appear to be highly dependent on the type of cannabinoid used, the dose and route of administration.

As more high quality trials are conducted, further evidence will be available on the use and effects of cannabis for neurological disorders.

References

1. Breivogel CS, Sim-Selley LJ. Basic neuroanatomy and neuropharmacology of cannabinoids. Int Rev Psychiatry. 2009;21(2):113–21.
2. Glass M, Dragunow M, Faull RL. Cannabinoid receptors in the human brain: a detailed anatomical and quantitative autoradiographic study in the fetal, neonatal and adult human brain. Neuroscience. 1997;77(2):299–318.
3. Zou M, et al. Role of the endocannabinoid system in neurological disorders. Int J Dev Neurosci. 2019;76:95–102.
4. den Boon FS, et al. Excitability of prefrontal cortical pyramidal neurons is modulated by activation of intracellular type-2 cannabinoid receptors. Proc Natl Acad Sci U S A. 2012;109(9):3534–9.
5. Zou S, Kumar U. Cannabinoid receptors and the endocannabinoid system: signaling and function in the central nervous system. Int J Mol Sci. 2018;19(3):833.
6. Bloomfield MAP, et al. The neuropsychopharmacology of cannabis: a review of human imaging studies. Pharmacol Ther. 2019;195:132–61.
7. Elliott J, et al. Cannabis-based products for pediatric epilepsy: an updated systematic review. Seizure. 2020;75:18–22.
8. Miller I, et al. Dose-ranging effect of adjunctive oral cannabidiol vs placebo on convulsive seizure frequency in Dravet syndrome: a randomized clinical trial. JAMA Neurol. 2020;77(5):613–21.
9. Devinsky O, et al. Trial of cannabidiol for drug-resistant seizures in the Dravet syndrome. N Engl J Med. 2017;376(21):2011–20.
10. Thiele EA, et al. Cannabidiol in patients with seizures associated with Lennox-Gastaut syndrome (GWPCARE4): a randomised, double-blind, placebo-controlled phase 3 trial. Lancet. 2018;391(10125):1085–96.
11. Devinsky O, et al. Randomized, dose-ranging safety trial of cannabidiol in Dravet syndrome. Neurology. 2018;90(14):e1204–11.
12. Cunha JM, et al. Chronic administration of cannabidiol to healthy volunteers and epileptic patients. Pharmacology. 1980;21(3):175–85.
13. Tzadok M, et al. CBD-enriched medical cannabis for intractable pediatric epilepsy: the current Israeli experience. Seizure. 2016;35:41–4.
14. McCoy B, et al. A prospective open-label trial of a CBD/THC cannabis oil in Dravet syndrome. Ann Clin Transl Neurol. 2018;5(9):1077–88.
15. Porcari GS, et al. Efficacy of artisanal preparations of cannabidiol for the treatment of epilepsy: practical experiences in a tertiary medical center. Epilepsy Behav. 2018;80:240–6.
16. Rog DJ, et al. Randomized, controlled trial of cannabis-based medicine in central pain in multiple sclerosis. Neurology. 2005;65(6):812–9.
17. Zajicek J, et al. Cannabinoids for treatment of spasticity and other symptoms related to multiple sclerosis (CAMS study): multicentre randomised placebo-controlled trial. Lancet. 2003;362(9395):1517–26.
18. Whiting PF, et al. Cannabinoids for medical use: a systematic review and meta-analysis. JAMA. 2015;313(24):2456–73.
19. Koppel BS, et al. Systematic review: efficacy and safety of medical marijuana in selected neurologic disorders: report of the Guideline Development Subcommittee of the American Academy of Neurology. Neurology. 2014;82(17):1556–63.

20. Nielsen S, et al. The use of cannabis and cannabinoids in treating symptoms of multiple sclerosis: a systematic review of reviews. Curr Neurol Neurosci Rep. 2018;18(2):8.
21. Carlini EA, Cunha JM. Hypnotic and antiepileptic effects of cannabidiol. J Clin Pharmacol. 1981;21(S1):417s–27s.
22. Nicholson AN, et al. Effect of Delta-9-tetrahydrocannabinol and cannabidiol on nocturnal sleep and early-morning behavior in young adults. J Clin Psychopharmacol. 2004;24(3):305–13.
23. Vigil JM, et al. Effectiveness of raw, natural medical cannabis flower for treating insomnia under naturalistic conditions. Medicines (Basel). 2018;5(3):75.
24. Chagas MH, et al. Cannabidiol can improve complex sleep-related behaviours associated with rapid eye movement sleep behaviour disorder in Parkinson's disease patients: a case series. J Clin Pharm Ther. 2014;39(5):564–6.

Cannabinoid Treatment for Rheumatic Disease

10

Mary-Ann Fitzcharles

Introduction

Rheumatic diseases are mostly lifelong, which essentially translates to prolonged treatment and a focus on symptom management. The objective in rheumatology clinical care is to attenuate the disease and symptoms and ensure that functional status of the patient is maintained. Given that many rheumatological conditions are associated with chronic pain, cannabis can be used as an adjunct given its opioid sparing properties [1]. The Canadian Rheumatology Association has developed a position statement for the use of medical cannabis for rheumatology patients [2]. This statement outlines the evidence of cannabis use related to efficacy, highlights cautions when cannabis is prescribed, and provides pragmatic advice regarding use.

Overview of Rheumatic Diseases

- Most people will experience musculoskeletal pain in their lifetime. Among the elderly population, OA is the most common condition that commonly results in pain and physical disability [3].
- The primary focus of treatment for any rheumatic condition is to identify and treat the underlying process, if possible, whilst attending to the principles of symptom relief.

M.-A. Fitzcharles (✉)
Alan Edwards Pain Management Unit, McGill University Health Centre, Montreal, QC, Canada

Division of Rheumatology, McGill University Health Centre, Montreal, QC, Canada

Montreal General Hospital, McGill University Health Centre, Montreal, QC, Canada
e-mail: Mary-Ann.Fitzcharles.med@ssss.gouv.qc.ca

R. Valani (ed.), *Cannabis Use in Medicine*,
https://doi.org/10.1007/978-3-031-12722-9_10

- The three broad categories of rheumatic diseases for which cannabis can be considered:
  - Inflammatory arthritis (IA),
  - Osteoarthritis (OA), and
  - Chronic widespread pain or fibromyalgia (FM).
- Soft tissue rheumatism, that includes tendonitis and bursitis, is usually self-limited and less often evolves into a chronic condition. Therefore, there is little use for cannabis in this setting.
- There is no "gold standard" regarding the ideal management of rheumatic pain, especially when chronic, but even small changes in pain can positively impact patient well-being.
  - Pain management will parallel with treatment of the underlying condition and should incorporate non-pharmacologic and pharmacologic treatments. See Chap. 11 on The use of cannabis for pain management when considering cannabis as an adjunct for pain control.
  - Cannabis-based medicines hold hope for symptom relief in rheumatic diseases. As we begin to understand more about the disease process and effects of the endocannabinoid system, the role of cannabis for rheumatological conditions will become clearer with further clinical studies.

Inflammatory Arthritis (IA)

- IA includes a wide spectrum of diseases such as rheumatoid arthritis (RA), juvenile inflammatory arthritis, seronegative arthritis (negative rheumatoid factor), psoriatic arthritis, inflammatory spondylarthritis and other less common connective tissue diseases.
- The main treatment objective is control of disease with disease-modifying antirheumatic drugs (DMARDs), either synthetic or biologic. Such treatments are usually continued indefinitely.
- In addition to modifying the disease process, symptom relief strategies are usually required in parallel with treatments.
  - Up to 50% of patients with IA will have persistent pain identified as "remaining pain", and up to 30% will have an associated FM.
  - Therefore, the ability to reduce pain and improve function can improve the quality of life of the patient.

Osteoarthritis (OA)

- OA can affect the small and large joints as well as the spine.
- It is the most prevalent form of arthritis affecting almost all persons in the later years of life, and is one of the most common causes of chronic pain [4].
- There is no DMARD that impacts the disease process of OA and treatments are focused on symptom relief.

Fibromyalgia (FM)

- FM most commonly presents in women in the 30–40-year range.
- FM is characterized by chronic diffuse body pain and important symptoms of sleep disturbance and fatigue. Many patients also experience mood disorders such as depression and anxiety.
- The pathophysiology of FM has underpinnings as a neurological dysfunction.
- The pain of FM is due to sensitization of the nervous system through a newly recognized mechanism known as nociplastic pain [5]. This is in contrast to nociceptive pain (caused by inflammation and tissue damage) or neuropathic pain (pain caused by nerve damage). See Chap. 11 on The use of cannabis for pain management for more details on the different types of pain.
- This is a difficult disease to manage, and symptoms are often poorly controlled with current treatments.
- FM can occur in association with other rheumatic diseases and greatly impacts quality of life.

The Endocannabinoid System and Rheumatological Conditions

- The endocannabinoid system promotes homeostasis and functions to return the body to equilibrium in response to a "fight and fly" stimulus.
- This system comprises at least two receptors (CB_1 and CB_2) and numerous endogenous ligands (endocannabinoids) that are produced on demand. See Chap. 3 on The Pharmacology of Cannabis.
- Synovial fluid analyses reveal that joints affected by IA or OA express cannabinoid receptors and produce endocannabinoids [6, 7]
  - Cannabinoid receptors are present in chondrocytes, synovium, and bone.
  - Increased levels of the endocannabinoids anandamide and 2-AG are seen in the synovium of patients with rheumatoid and osteoarthritis.
- Upregulation of the endocannabinoid system is seen as a response to the inflammatory stimulus and functions to reduce hypersensitivity of joint nociceptors and therefore attenuate pain. See Fig. 10.1.
  - The cannabinoid system decreases inflammatory mediators within the synovium.
  - The endocannabinoid system is an important pathway in pain modulation. See Chap. 11 on the use of cannabis for pain management.
  - CB_1 agonists as well as inhibitors of fatty acid amide hydrolase (normally degrades anandamide) and monoacylglycerol lipase (normally degrades 2-AG) are shown to decrease nociceptive pain.
- The nervous system effects of the endocannabinoid system reduce both peripheral and central nervous system sensitization. This is the physiological mechanism that is operative in nociplastic pain conditions such as FM and "remaining pain" in patients with IA.

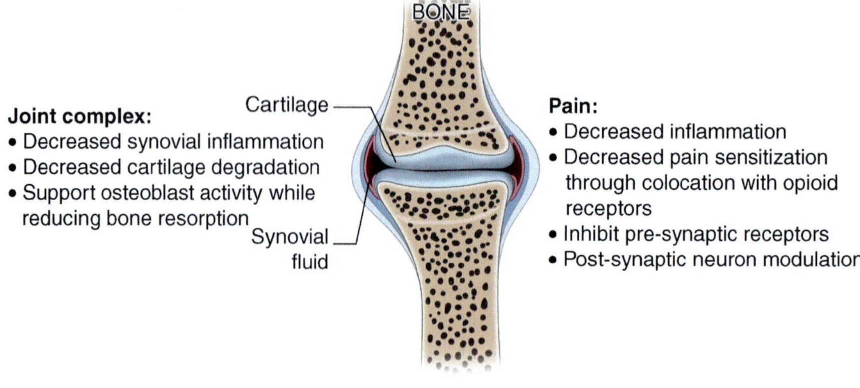

Fig. 10.1 Potential effects of cannabis in the synovial joint

The Role of Cannabis in Rheumatological Conditions

- Preclinical studies have shown attenuation of inflammation, joint destruction and pain in animal models of IA and inflammation [8, 9].
- There is a paucity of well-designed clinical trials to show the benefits of cannabis in patients with rheumatological conditions.
  - To date, less than 200 patients with rheumatic diseases have been studied in randomized clinical trials (RCTs), with not a single study for the effect of medical cannabis (the herbal product) [10].
- Pain, which is a major symptom of many rheumatic diseases, is associated with sleep disturbance and impacts mood. Both of these are affected by cannabinoids.
  - Given that current pain medications are suboptimal with considerable side effects, cannabis is a promising alternative for symptom management.
  - Easier accessibility of the product has allowed patients the opportunity to experiment with use and often self-medicate. With ongoing research, there may be the potential to use cannabis in parallel with current approved therapy of rheumatic conditions.

Key Points for Medical Cannabis Use in Rheumatology Patients

- Rheumatic diseases are mostly lifelong, therefore when a treatment is initiated and is successful, it can be anticipated that use of that therapy will be prolonged.
  - The objective in rheumatology clinical care is to attenuate the disease and symptoms and ensure that function is maintained.
  - The Canadian Rheumatology Association has developed a position statement for the use of medical cannabis for rheumatology patients [2]. This statement

outlines the evidence for effect, highlights the cautions for use, and provides pragmatic advice regarding use.

Medical cannabis is not an alternative to standard care for any rheumatic disease.

When patients seek advice about medical cannabis, physicians must remain empathetic, avoid personal biases, and ensure safety for patients and society.

There are no clinical trials of medical cannabis in rheumatology patients. Current evidence for effect of cannabis-based medicines in fibromyalgia, osteoarthritis, rheumatoid arthritis, and back pain is limited.

Extrapolating from other conditions, medical cannabis may provide some symptom relief in some patients.

Short-term side effects such as dizziness, appetite changes and effect on mood can be expected. Psychomotor effects can be anticipated up to 5 h after use of inhaled THC, and may persist for up to 24 h [11].

The long-term risks in patients with rheumatic diseases are unknown.

Treatment Principles to Reduce Harm

- Before starting cannabis for any patient, the health care provider should provide appropriate information and set realistic outcomes. (See Chap. 4 for further details on patient assessment and dosing guidelines)
  - An assessment for substance use disorder must be documented.
  - Patients should be well informed about the adverse effects associated with cannabis use.
  - Oral administration is the preferred route. Cannabis should not be smoked as smoking is associated with increased severity of IA. Vaping is not benign and has been associated with severe lung disease [12].
  - Cannabis with a low THC content (maximum 9%) and higher CBD content is preferable.
  - Total daily dose should not exceed 3 g, either regularly or on demand.
  - Follow-up should be within 4–8 weeks after initiating the drug, and thereafter at 3 monthly intervals.

Contraindications and Cautions for Medical Cannabis

- Medical cannabis should not be used in the following patients:
  - Rheumatology patients under the age of 25 years.
  - Patients with allergic reactions to cannabinoid products.
  - Women who are pregnant or breastfeeding.
  - Patients with a history of a psychotic illness, substance abuse disorder, previous suicide attempts or suicidal ideation.

- Medical cannabis should be used with caution in the following patients:
  - Patients with unstable mental health disease.
  - Patients with current moderate or severe cardiovascular or pulmonary disease.
  - Patients working in settings requiring high levels of cognitive ability, concentration, and alertness.
  - Patients receiving concomitant therapy with sedative-hypnotics or other psychoactive drugs.

The Adolescent and Young Adult with a Rheumatic Disease

- As cannabis is gaining mainstream acceptance and is legalized for recreational use in many jurisdictions, there is a prevalent perception of safety. This applies particularly for young people.
  - Knowledge of the adverse effects in young persons is only available from recreational users without any data regarding this patient population. There are specific cautions for young people.
    - The endocannabinoid system influences development and maturation of the nervous system, which continues into the early 20s [13].
    - Cannabis in adolescents is associated with reduced educational and lifetime achievement and is a risk of mental health disease [14, 15].
    - Cannabis is an addictive substance, with increased rates of addiction related to early age of onset of use, and regular use. Rates are reported as 9% of all users, 17% for onset of use in adolescence and 25–50% for daily users [16].

The Adult with a Rheumatic Disease

- This is the time of life when persons are establishing careers, have family and social responsibility, but also the most common time for onset for IA and FM.
  - In this context patients will need to remain functional, which means work and safety issues as well as operating a vehicle. See Chap. 13 on the occupational considerations related to cannabis use.
  - Care should also be taken for women of childbearing age.
    - Cannabis crosses the placenta and should not be used in pregnancy. There is evidence for reduced birthweight, preterm birth and requirement for neonatal intensive care [17]. See Chap. 15 on Cannabis use in the pregnant patient.
    - Driving after cannabis use is associated with psychomotor impairment that has been measured up to 5 h in healthy adults and represents twice the risk of a motor vehicle accident [11].
      - Given the pharmacokinetic, pharmacodynamic, and metabolic complexities with cannabis use, a simple dose dependent effect is difficult to elicit with cannabis. See Chap. 13 on the occupational considerations related to cannabis use.

The Elderly Person with a Rheumatic Disease

- Many elderly patients may wish to explore cannabis use for pain management and help with sleep. In the elderly person with OA, use of cannabis should be cautioned for the following reasons:
  - Psychomotor effects which increase the risk for falls.
  - These effects will be compounded in the presence of other agents that have psychoactive effects.
  - Cannabis use is a risk for acute cardiovascular and cerebral events [18].

A Pragmatic Approach to Cannabis Recommendations

- Understand the reasons why a patient wishes to use cannabis.
- Evaluate previous therapeutic trials, including a possible treatment trial of a pharmaceutical cannabis-based medicine.
- The physician should be fully knowledgeable of the patient and must remain responsible for patient care.
- Prior to any recommendation for use, there must be a comprehensive clinical evaluation, including a psychosocial and mental health history.
- The physician and patient must agree to the goals of therapy, which should include both symptom relief and maintenance of function.
- Treatment should be initiated slowly, beginning with a nighttime dose, and not exceeding 3 g a day, the maximum average dose recommended by Health Canada [19].
- Cannabis should be obtained legally from a registered grower and not via the illegal route.
- Cannabis with a low THC content and higher CBD content is preferable. Studies have used THC content up to 12.5%, but with a high rate of adverse events at this concentration [20].
- The ideal dosing schedule is unknown, with no dose finding studies to examine optimal daily amount or specific molecular concentrations of THC and CBD [21].
- Medical cannabis is ideally not a lifetime treatment, and at each visit justification for continued treatment must be documented.

There remains much uncertainty regarding medical cannabis in rheumatic disease. Physicians have a responsibility to ensure that patients are fully informed regarding the current evidence and recommendations around the use of cannabis as a therapeutic agent. For those patients who are self-medicating with cannabis, physicians must make efforts to ensure harm reduction through education. Whether medical cannabis will finally emerge as a standard treatment option for rheumatology patients remains to be seen in the years to come.

References

1. Hulland O, Oswald J. Cannabinoids and pain: the highs and lows. Rheum Dis Clin N Am. 2021;47(2):265–75.
2. Fitzcharles MA, et al. Position statement: a pragmatic approach for medical cannabis and patients with rheumatic diseases. J Rheumatol. 2019;46(5):532–8.
3. Helmick CG, et al. Estimates of the prevalence of arthritis and other rheumatic conditions in the United States. Part I. Arthritis Rheum. 2008;58(1):15–25.
4. Malfait AM, Miller RE, Miller RJ. Basic mechanisms of pain in osteoarthritis: experimental observations and new perspectives. Rheum Dis Clin N Am. 2021;47(2):165–80.
5. Fitzcharles MA, et al. Nociplastic pain: towards an understanding of prevalent pain conditions. Lancet. 2021;397(10289):2098–110.
6. Richardson D, et al. Characterisation of the cannabinoid receptor system in synovial tissue and fluid in patients with osteoarthritis and rheumatoid arthritis. Arthritis Res Ther. 2008;10(2):R43.
7. Sarzi-Puttini P, et al. Cannabinoids in the treatment of rheumatic diseases: pros and cons. Autoimmun Rev. 2019;18(12):102409.
8. Barrie N, Manolios N. The endocannabinoid system in pain and inflammation: Its relevance to rheumatic disease. Eur J Rheumatol. 2017;4(3):210–8.
9. Katz-Talmor D, et al. Cannabinoids for the treatment of rheumatic diseases—where do we stand? Nat Rev Rheumatol. 2018;14(8):488–98.
10. Fitzcharles MA, et al. Efficacy, tolerability and safety of cannabinoids in chronic pain associated with rheumatic diseases (fibromyalgia syndrome, back pain, osteoarthritis, rheumatoid arthritis): a systematic review of randomized controlled trials. Schmerz. 2016;30(1):47–61.
11. Ogourtsova T, et al. Cannabis use and driving-related performance in young recreational users: a within-subject randomized clinical trial. CMAJ Open. 2018;6(4):E453–62.
12. Hammond D. Outbreak of pulmonary diseases linked to vaping. BMJ. 2019;366:l5445.
13. Harkany T, et al. Endocannabinoid functions controlling neuronal specification during brain development. Mol Cell Endocrinol. 2008;286(1–2 Suppl 1):S84–90.
14. Meier MH, et al. Persistent cannabis users show neuropsychological decline from childhood to midlife. Proc Natl Acad Sci U S A. 2012;109(40):E2657–64.
15. Jager G, Ramsey NF. Long-term consequences of adolescent cannabis exposure on the development of cognition, brain structure and function: an overview of animal and human research. Curr Drug Abuse Rev. 2008;1(2):114–23.
16. Curran HV, et al. Keep off the grass? Cannabis, cognition and addiction. Nat Rev Neurosci. 2016;17(5):293–306.
17. Gunn JK, et al. Prenatal exposure to cannabis and maternal and child health outcomes: a systematic review and meta-analysis. BMJ Open. 2016;6(4):e009986.
18. Singh A, et al. Cardiovascular complications of marijuana and related substances: a review. Cardiol Ther. 2018;7(1):45–59.
19. Health Canada, Information for Health Care Professionals—Cannabis and the Cannabinoids, Health Canada Controlled Substances and Tobacco Directorate, Editor. 2013.
20. Ware MA, et al. Cannabis for the management of pain: assessment of safety study (COMPASS). J Pain. 2015;16(12):1233–42.
21. Allan GM, et al. Simplified guideline for prescribing medical cannabinoids in primary care. Can Fam Physician. 2018;64(2):111–20.

The Use of Cannabis for Pain Management

11

Rahim Valani

Introduction

Chronic pain affects up to 40% of the population, and a significant number describe their pain as moderate or severe. This has considerable impact at an individual and societal level. These individuals are in constant pain and are limited in their personal and occupational activities. Treatment strategies using both pharmacological and non-pharmacological options are recommended for chronic pain. The National Academies of Science, Engineering, and Medicine indicated that there is substantial evidence for the use of cannabis for chronic pain in adult patients [1]. With increasing evidence on the efficacy of cannabis, the ability to reduce opioid use, as well as patient preference, clinicians must be versed in the use of cannabis for chronic pain.

Chronic Pain

- Chronic pain affects 11–40% of the population and carries with it a significant burden of disease.
  - It affects all facets of the patient's life including their personal, social, and occupational abilities.
  - Pain is the most common reason for adults to seek medical care. It is also the most prevalent symptom to seek medical care among elderly patients [2–4].
    11% of chronic pain patients rate their pain as being severe pain.
    12% of chronic pain patients are affected to the point of being classified as disabled.

R. Valani (✉)
Department of Medicine, University of Toronto, Toronto, ON, Canada
e-mail: r.valani@utoronto.ca

© Springer Nature Switzerland AG 2022
R. Valani (ed.), *Cannabis Use in Medicine*,
https://doi.org/10.1007/978-3-031-12722-9_11

- Chronic pain is defined as pain that persists beyond the expected healing time. While acute pain usually stems from injury or information, it usually subsides over time. If the pain persists over 3–6 months, then it is classified as chronic pain.
- The International Association for the Study of Pain (IASP) further classifies chronic pain into the following categories: [5]
  - Chronic primary pain.
    This is defined as pain that persists longer than 3 months and is associated with significant emotional distress or functional disability.
  - Chronic cancer related pain.
  - Chronic post-surgical or post-traumatic pain.
  - Chronic neuropathic pain.
  - Chronic secondary headache or orofacial pain.
  - Chronic secondary visceral pain.
  - Chronic secondary musculoskeletal pain.
- Inadequate pain management has been linked to:
  - Reduced quality of life from ongoing pain as well as limitations in activity.
  - Poor sleep.
  - Impaired physical function.
  - Poor mental health (this can range from low mood to major depression and low self esteem).
  - Physiological consequences from poor pain management.
- There are social, demographic, and clinical factors associated with chronic pain. These include: [3, 4, 6]
  - Demographic variables related to chronic pain:
    Female gender.
    Older age.
    Lower socio-economic status.
    Geographical location.
    Cultural background.
    Employment status.
  - Lifestyle factors that contribute to chronic pain:
    Smoking or alcohol use.
    Physical inactivity.
    Poor nutrition.
  - Clinical variables associated with chronic pain include:
    Mental health issues.
    Being overweight.
    Sleep disorder.
    Genetic predisposition.
  - Other factors:
    History of abuse / inter-personal violence.
    Attitudes and belief about pain.
- Management of chronic pain is a major challenge given the complexity of the symptom itself.
  - Chronic pain should be managed in a multi-modal fashion that limits the amount of opioid medications used.

Table 11.1 Summary of the categories of pain

	Nociceptive	Neuropathic	Nociplastic
Mechanism	Direct damage of tissue that stimulate nociceptors.	From direct nerve damage.	Alteration of nociception with no evidence of tissue damage or somatosensory system causing pain.
Physical cause	Injury or inflammation.	Nerve damage from damage, compression, or irritation.	CNS dysfunction.
Example	Direct trauma / injury Surgery	Diabetic neuropathy Shingles	Fibromyalgia Chronic widespread pain

CNS Central nervous system

Pathophysiology of Pain

- Pain is defined as an unpleasant sensation in response to real or perceived tissue damage.
- Pain has been classified into three main categories: (See Table 11.1)
  - Nociceptive pain.
    - This type of pain is caused by direct stimulation of nociceptors that are sensitive to noxious physical stimuli.
    - The receptors relay impulses through the somatosensory system to the spinal cord and then thalamus and cerebral cortex.
    - There are two main types of nerve fibres involved:
      - $A\delta$ nerve fibres which are are thin, myelinated fibres. These fibres respond to mechanical and thermal stimuli.
      - C fibres, which are unmyelinated, respond to a variety of stimuli.
  - Neuropathic pain.
    - This pain sensation is caused by damage to nerves in either the central or peripheral nervous system.
    - The resultant nerve damage causes abnormal activity of impulses triggering the pain.
    - There is a lack of correlation between the actual stimulus and perceived pain sensation.
  - Nociplastic pain.
    - This is due to an alternation in nociception without any apparent cause that triggered the activation of nociceptors or cause of injury.

The Endocannabinoid System and Pain

- CB_1 receptors are primarily located in the central nervous system, while CB_2 receptors are expressed in immune cells. See Chap. 3 on Pharmacology of cannabis.

- There have been several mechanisms of action suggested for the analgesic effects of cannabinoids: [7, 8]
    - CB_1 receptors and the mu-opioid receptors are distributed in many of the same areas in the brain.
        The bidirectional relationship between the opioid mu-receptor and CB_1 receptor have similar reward properties with drugs of misuse.
    - Inhibit presynaptic neurotransmitter release.
        CB_1 receptors located in the afferent neurons are the main target for cannabinoid analgesia.
    - Post-synaptic neuron modulation.
    - Decreases inflammation.
    - Nerve injury causes an increase in CB_1 receptors in the dorsal root ganglion, spinal cord, and other brain areas related with the emotional component of pain [8, 9].
    - Orphan receptors GPR18 and GPR55 have been suggested as cannabinoid receptors and may be involved with sensory transmission and pain regulation [9].

Chronic Pain Assessment and Management Strategies

- Appropriate chronic pain management requires a holistic approach to the patient to not only understand the patient's symptoms, but also factors that can contribute to poor quality of life.
    - This strategy includes having a multi-modal approach to chronic pain. See Fig. 11.1.
- Pain management usually includes analgesic agents. The main classes of agents used are:
    - Non-opioid analgesia.
        Non-steroidal anti-inflammatory agents (NSAIDS), which are cyclooxygenase enzyme inhibitors, are the main agents under this class.
    - Local anesthetics which are sodium channel blockers.
    - Antidepressants which are serotonin (5-HT) reuptake inhibitors or noradrenaline reuptake inhibitors.
    - Anticonvulsants like gabapentin and pregabalin decrease synaptic transmission.
    - Opioid analgesia (agents such as morphine or hydromorphone).
        These agents work through specific opioid receptors (mu, delta, kappa). All these receptors have analgesic effects.
        - Mu (OP3, MOP) – responsible for dopamine dependent reward and reinforcement for opioids.
        - Delta (OP1, DOP) – is involved with physical dependence.
        - Kappa (OP2, KOP) – blocks the rewarding effects of mu agonists.

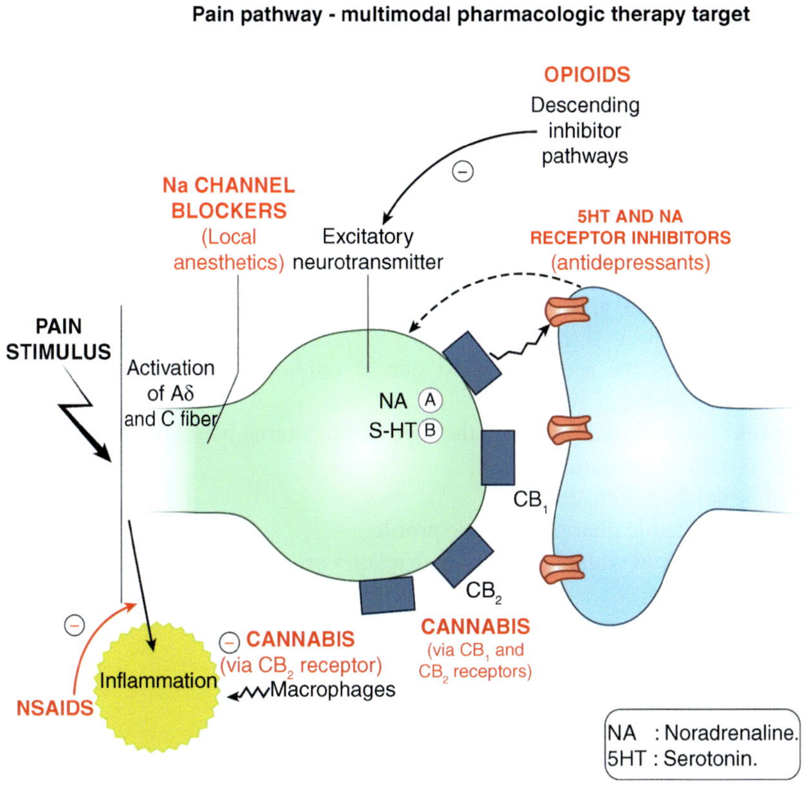

Fig. 11.1 Multimodal pharmacological management of pain

Opioid Therapy for Chronic Pain

- Opioids are commonly used for chronic pain. However, it requires individual dose titration [2, 10].
  - For elderly patients in whom chronic pain is highly prevalent, care should be taken with the use of opioids. Issue concerning opioid use in this population are as follows:

    This population is already at risk of cognitive impairment. Adding opioids can increase confusion states and result in impaired judgement.

    Physiological changes that alter the pharmacokinetics of opioids require lower doses and slower titration.

    Polypharmacy is common in the elderly and requires vigilant review for drug-drug interactions.

    Common side effects of opioids in the elderly include constipation, vomiting, urinary retention, and sedation.

- Opioid misuse and abuse are common [11, 12].
  - 4% of the adult US population misuse opioid prescriptions.
  - Diversion and misuse of opioids along with the availability of illicit heroin and fentanyl have culminated in the opioid epidemic.
  - Establishing a prior history of substance abuse is a strong predictor of opioid misuse / abuse, so care must be taken in prescribing opioids to this population [13]. Strict adherence to protocols and contracts with the patient can help reduce opioid misuse.
  - The US has the highest growth in morbidity and mortality associated with opioids. Since 2014, drug overdose has been the leading cause of accidental death in the US [14].
    - Canada has the second highest rate of opioid prescription and is also affected by the crisis.
- The opioid epidemic has prompted the need to find alternatives that are safer. The goal is to find a substitute that: [15]
  - Can be safely administered.
  - Has a predictable pharmacokinetic profile.
  - Can be produced with high purity with quality processing methods that do not use toxins.
  - Has a good safety profile.

Cannabis Use for Opioid Detoxification

- Cannabis has also been suggested as an adjunct to standard opioid detoxification programs to reduce withdrawal symptoms [16].
  - Withdrawal from heroin usually starts 4–6 h after discontinuation, whereas longer acting opioids peak at 24–48 h. Symptoms can last up to 2 weeks.
  - Symptoms of opioid withdrawal are usually managed by:
    - Mu receptor agonists (methadone, buprenorphine).
    - α-adrenergic agonists (clonidinie, lofexidine).
    - NSAIDs or acetaminophen for pain.
    - Loperamide for diarrhea.
    - Ondansetron for nausea.
    - Trazodone for insomnia.
  - Cannabis can counteract the withdrawal symptoms as described in Table 11.2:

Table 11.2 Benefits of cannabis in opioid use withdrawal symptoms

Opioid withdrawal symptom	Cannabis counter-effects
Anxiety	Anxiolysis and sedation
Hypertension, tachycardia	Hypotension
Nausea, vomiting, abdominal cramps	Anti-emetic
Tremors	Decreased tremors
Myalgias, arthralgias	Analgesia, anti-inflammatory

- CB$_1$ receptors are involved with dopamine regulation. These receptors are co-localized in the nucleus accumbens and dorsal striatum, which are areas involved with reward, goal-direction, and habit. Hence, the shared response.

Use of Cannabis for Chronic Pain

- It is estimated that 15% of patients suffering from pain in Australia and Canada self medicate with cannabis. The most common cited reason for cannabis use among these patients is for chronic pain [1, 17].
- The analgesic effects of cannabinoids occur through the following mechanisms: [18]
  - Interacting with G-protein coupled receptor (GPR55).
  - Opioid receptors.
  - Serotonin receptors
  - α 3 glycine receptors.
  - At the spinal level, cannabinoids exert their analgesic effects by activation if the kappa (OP2) opioid receptor [19].
- Cannabis is not considered a first line agent in the management of chronic pain. It has, however, been shown to be useful as an adjunct [20].
  - Current evidence points to the potential to use cannabis as an opioid sparing agent that also has the capability of synergistic analgesia [7, 21–23].
    Cannabis as an adjunct therapy has the potential to decrease opioid use in chronic pain patients by 40–60% [7, 24–27].
    Retrospective study showed a decrease of 40 mg of morphine equivalents in 6 months with use of cannabis [22].
    When cannabis is used in conjunction with morphine, the plasma concentration of the opioid is unaffected.
- The evidence shows that jurisdictions in which cannabis use has been legalized has resulted in a 23% decrease in hospitalizations from opioid dependence and abuse, as well as a 13% decrease in opioid overdose [23].
- It is difficult to compare studies on the efficacy and utility of cannabis for chronic pain due to the heterogeneity of patients, the different chemical variety of cannabis used, and definitions used to classify which patients are being studied. See Chap. 6 on Evidence based reviews on cannabis therapy.
  - Not surprisingly, there is contradictory evidence supporting the use of cannabis for pain in general. Much of this stems from: [15, 28]
    A lack standard dosing of cannabis products. The lack of standard dosing, variability in CBD/THC components, and routes of administration makes comparison difficult.
    There are no large, well designed randomized controlled trials.
    - Small sample size.
    - Heterogenous population.
    - Variable treatment duration.
    - Appropriate monitoring of therapy.

The pain scores used in these studies may not be validated for this subset of patients.

Most studies use cannabis as an adjunct rather than a primary agent.

While there is ample subjective evidence of improvement in pain, strong objective data is lacking.

There are inherent biases among the medical and legal community for the use of cannabis that stems form poor knowledge. This is compounded by the lack of good clinical trials which is difficult to conduct given the regulatory requirements. See Chap. 7 on Evidence based reviews on cannabis therapy.

- Conflicting regulatory frameworks. For example, the State versus Federal laws in the US are conflicting in certain areas. This makes it difficult for health care professionals, patients, and law enforcement agencies alike to comprehend what is legal.
- Political and public opinion on the use of cannabis based on the stigma against cannabis use.

A lack of evidence should not be mistaken for a lack of benefit / efficacy.

Ignoring the "entourage effect"—the combination of all the molecules in the cannabis plant has better efficacy than isolates. Taking the components of the cannabis plant in isolation may not produce the same effect as with the whole plant use.

- Unfortunately, the majority of guidelines and recommendations have not addressed patient perceptions, shared decision making, and safe effective treatment for the use of cannabis.

- Chronic pain patients who have used cannabis tend to be exposed to high levels of THC, which results in more side effects from activation of CB_1 receptors.
  - For chronic pain, patients could achieve pain management with THC concentrations between 10-15% [15].
- Cannabis has been shown to be useful in the management of neuropathic pain (peripheral neuropathy). There has been good evidence in benefits of cannabis in patients with neuropathies secondary to chemotherapy, diabetes, HIV, cancer, and fibromyalgia.
  - The CB_2 receptor is involved in modulating inflammation. CB_2 agonists could therefore be an important treatment pathway for neuropathic pain.
  - Cannabis has been shown to be cost effective when used as an adjunct treatment strategy for patients with peripheral neuropathy [29].
  - According to a recent Cochrane review, the benefits of cannabis in chronic neuropathic pain may outweigh the harms caused by using cannabis [30].
  - There are ongoing studies to determine its use in nociplastic pain, specifically with rheumatologic conditions [21, 31]. See Chap. 10 for Cannabinoid treatment for rheumatic disease.
- Patients with pain also have associated symptoms, and cannabis has been shown to be of benefit in reducing these symptoms.

Table 11.3 Assessment of patients for starting and continuing cannabis for chronic pain

Questions to consider for cannabis therapy	
What is the indication for prescribing cannabis?	Understand why the patient would benefit from cannabis use.
Are there alternative treatments available?	While cannabis is helpful for many conditions, including pain, many practitioners still use this as adjunct therapy. Knowing the alternatives is an important discussion item for shared decision making.
What are the risks / side effects?	Knowing the side effects and risks of cannabis as well as the alternative treatment strategies is important. Explain all of these to the patient.
Treatment	Start treatment as recommended and titrate up slowly.
Reassessment	Regular follow up is necessary to know the benefits of cannabis treatment and to monitor any adverse effects.

- Among cancer patients, the use of cannabis has been shown to: [32, 33]
    Improved sleep.
    Decreased fatigue.
    Lower anxiety/depression.
    Less nausea/vomiting.
- For all patient with chronic pain, cannabis has been shown to: [34]
    Reduce pain (moderate level of evidence).
    Improve physician function (high level of evidence).
- A recent systematic review and meta-analysis from 2021 concluded that non-inhaled cannabis resulted in a small to very small improvement in pain relief, physical functioning, and sleep in patients with chronic pain [35].
- There is the potential for opioids and cannabis abuse, and this requires appropriate monitoring of patients (Table 11.3). See Chap. 4 on Patient assessment and dosing recommendations for cannabis. See Table 11.1 on assessment of patients.
    - At each visit, it is important to review and document:
        Analgesic effects.
        Activity levels including activities of daily living and if these can be carried out independently.
        Aberrant behaviour.
        Any adverse effects which may necessitate a change in dosing.

Summary

The evidence is growing for the use of Cannabis in patients with chronic pain, and specifically neuropathic pain. Ongoing research and a better understanding of the endocannabinoid system's role in inflammation and analgesia could potentially allow cannabis to be used for nociplastic pain syndromes. Furthermore, as an opioid sparing agent it has the ability to decrease the use of opioids in high-risk individuals (elderly patients, those at risk for abuse / misuse).

References

1. National Academies of Sciences, E., and Medicine. The health effects of cannabis and cannabinoids: the current state of evidence and recommendations for research. Washingotn, DC: The National Academies Press; 2017.
2. Pergolizzi J, et al. Opioids and the management of chronic severe pain in the elderly: consensus statement of an International Expert Panel with focus on the six clinically most often used World Health Organization Step III opioids (buprenorphine, fentanyl, hydromorphone, methadone, morphine, oxycodone). Pain Pract. 2008;8(4):287–313.
3. Mills SEE, Nicolson KP, Smith BH. Chronic pain: a review of its epidemiology and associated factors in population-based studies. Br J Anaesth. 2019;123(2):e273–83.
4. Dahlhamer J, et al. Prevalence of chronic pain and high-impact chronic pain among adults—United States, 2016. MMWR Morb Mortal Wkly Rep. 2018;67(36):1001–6.
5. Treede RD, et al. Chronic pain as a symptom or a disease: the IASP classification of chronic pain for the international classification of diseases (ICD-11). Pain. 2019;160(1):19–27.
6. van Hecke O, Torrance N, Smith BH. Chronic pain epidemiology and its clinical relevance. Br J Anaesth. 2013;111(1):13–8.
7. Wiese B, Wilson-Poe AR. Emerging evidence for Cannabis' role in opioid use disorder. Cannabis Cannabinoid Res. 2018;3(1):179–89.
8. Vučković S, et al. Cannabinoids and pain: new insights from old molecules. Front Pharmacol. 2018;9:1259.
9. Guerrero-Alba R, et al. Some prospective alternatives for treating pain: the endocannabinoid system and its putative receptors GPR18 and GPR55. Front Pharmacol. 2018;9:1496.
10. Ravi D, et al. Associations between marijuana use and cardiovascular risk factors and outcomes: a systematic review. Ann Intern Med. 2018;168(3):187–94.
11. Skolnick P. The opioid epidemic: crisis and solutions. Annu Rev Pharmacol Toxicol. 2018;58:143–59.
12. Shipton EA, Shipton EE, Shipton AJ. A review of the opioid epidemic: what do we do about it? Pain Ther. 2018;7(1):23–36.
13. Turk DC, Swanson KS, Gatchel RJ. Predicting opioid misuse by chronic pain patients: a systematic review and literature synthesis. Clin J Pain. 2008;24(6):497–508.
14. Centers for Disease Control and Prevention. *Injuries and violence are leading causes of death.* 2021.; https://www.cdc.gov/injury/wisqars/animated-leading-causes.html.
15. Romero-Sandoval EA, Kolano AL, Alvarado-Vázquez PA. Cannabis and cannabinoids for chronic pain. Curr Rheumatol Rep. 2017;19(11):67.
16. Kudrich C, et al. Adjunctive management of opioid withdrawal with the nonopioid medication cannabidiol. Cannabis Cannabinoid Res. 2021;
17. Park JY, Wu LT. Prevalence, reasons, perceived effects, and correlates of medical marijuana use: a review. Drug Alcohol Depend. 2017;177:1–13.
18. Bennici A, et al. Safety of medical cannabis in neuropathic chronic pain management. Molecules. 2021;26:20.
19. Stella B, et al. Cannabinoid formulations and delivery systems: current and future options to treat pain. Drugs. 2021;81(13):1513–57.
20. Häuser W, et al. European Pain Federation (EFIC) position paper on appropriate use of cannabis-based medicines and medical cannabis for chronic pain management. Eur J Pain. 2018;22(9):1547–64.
21. Lossignol D. Cannabinoids: a new approach for pain control? Curr Opin Oncol. 2019;31(4):275–9.

22. O'Connell M, et al. Medical cannabis: effects on opioid and benzodiazepine requirements for pain control. Ann Pharmacother. 2019;53(11):1081–6.
23. Shi Y. Medical marijuana policies and hospitalizations related to marijuana and opioid pain reliever. Drug Alcohol Depend. 2017;173:144–50.
24. Sagy I, et al. Safety and efficacy of medical cannabis in fibromyalgia. J Clin Med. 2019;8:6.
25. Shah A, et al. Impact of medical marijuana legalization on opioid use, chronic opioid use, and high-risk opioid use. J Gen Intern Med. 2019;34(8):1419–26.
26. Bachhuber M, Arnsten JH, Wurm G. Use of cannabis to relieve pain and promote sleep by customers at an adult use dispensary. J Psychoactive Drugs. 2019;51(5):400–4.
27. Khan SP, Pickens TA, Berlau DJ. Perspectives on cannabis as a substitute for opioid analgesics. Pain Manag. 2019;9(2):191–203.
28. Campbell G, Stockings E, Nielsen S. Understanding the evidence for medical cannabis and cannabis-based medicines for the treatment of chronic non-cancer pain. Eur Arch Psychiatry Clin Neurosci. 2019;269(1):135–44.
29. Tyree GA, et al. A cost-effectiveness model for adjunctive smoked cannabis in the treatment of chronic neuropathic pain. Cannabis Cannabinoid Res. 2019;4(1):62–72.
30. Mücke M, et al. Cannabis-based medicines for chronic neuropathic pain in adults. Cochrane Database Syst Rev. 2018;3(3):Cd012182.
31. Modesto-Lowe V, Bojka R, Alvarado C. Cannabis for peripheral neuropathy: the good, the bad, and the unknown. Cleve Clin J Med. 2018;85(12):943–9.
32. Bar-Lev Schleider L, et al. Prospective analysis of safety and efficacy of medical cannabis in large unselected population of patients with cancer. Eur J Intern Med. 2018;49:37–43.
33. Anderson SP, et al. Impact of medical cannabis on patient-reported symptoms for patients with cancer enrolled in Minnesota's Medical Cannabis Program. J Oncol Pract. 2019;15(4):e338–45.
34. Busse JW, et al. Medical cannabis or cannabinoids for chronic pain: a clinical practice guideline. BMJ. 2021;374:n2040.
35. Wang L, et al. Medical cannabis or cannabinoids for chronic non-cancer and cancer related pain: a systematic review and meta-analysis of randomised clinical trials. BMJ. 2021;374:n1034.

Complications and Adverse Events from Cannabis Use

12

Anne Finlayson and Wesley Palatnick

Introduction

Cannabis is the most used illicit drug in the world. In the United States, as of 2019, cannabis is legal for medical purposes in 33 states and for recreational use in 11 states [1]. However, cannabis use remains illegal federally in America. In contrast, Canada and Uruguay are the two countries where the sale and consumption of recreational cannabis is legal (See Chap. 2 on the legal aspects of cannabis). Adult recreational use of cannabis became legal in 2018 in Canada with the passage of the *Cannabis Act*, and it is estimated that approximately 18% of the adult population use cannabis.

Although cannabis growth, sale and use in Canada is legal, there are still illegal operations that produce synthetic cannabinoids that are adulterated and of higher potency. The variability in concentration of cannabinoids and unknown adulterants used in the production of synthetic compounds increases the risk of adverse events. Recognizing the harmful effects is important for the medical practitioner to counsel the patient and provide appropriate care.

Cannabis in the Emergency Department

- Cannabis is known to have psychoactive effects. There are two major active chemical compounds that are well studied for their effects (See Chap. 3 on the Pharmacology of Cannabis):
  - Tetrahydrocannabinol (THC) is the major psychoactive component.
  - Cannabidiol (CBD) is attributed with having anti-inflammatory and analgesic properties.

A. Finlayson (✉) · W. Palatnick
Department of Emergency Medicine, University of Manitoba, Winnipeg, MB, Canada
e-mail: WPalatnick@wrha.mb.ca

© Springer Nature Switzerland AG 2022
R. Valani (ed.), *Cannabis Use in Medicine*,
https://doi.org/10.1007/978-3-031-12722-9_12

- There are three main types of *Cannabis* plants:
  - *Cannabis Sativa* – this is grown throughout the world and is the common type of cannabis plant seen in North America.
  - *Cannabis Indica* (skunk weed) – this is a short plant that has high concentrations of THC.
  - *Cannabis Ruderalis* – this type is primarily found in Central Asia.
- Of the different species, the *Indica* species is said to have more sedating effects whereas *Cannabis Sativa* is said to have more energizing effects.
- Cannabis is available in a variety of forms:
  - Marijuana (this is the dried flower of the *Cannabis Sativa* plant).
    Alternative and street names are weed, herb, pot, grass, bud, ganja, and Mary Jane.
  - Hashish or hash [2].
    This refers to the dried resin from the flowering tops of the unpollinated female cannabis plant.
    It is also known as Dab and Budder.
    - These compounds can contain greater than 70% THC.
  - Hash oil.
    This is also referred to as Shatter, Honey oil, or Cannabis oil.
    Hash oil is a concentrate of cannabis extract and contains high concentrations of THC.
    The compound can be smoked, vaporized, eaten, or applied topically.
  - Edibles.
    Edibles, as the name implies, are food products that contain cannabis and produce psychoactive effects.
- Cannabis can be consumed in a variety of ways, and the route of exposure determines the onset and duration of symptoms (See Chap. 3 on the Pharmacology of Cannabis):
  - Inhaled
    Onset is usually within 5–10 min with a duration of up to 4 h.
    Smoking and vaporization ("Vaping") are the two main modes.
  - Oral ingestion
    Onset is highly variable from 30 min–3 h, and the duration of action can last up to 12 h.
  - Topical application
    Cannabinoids are not well absorbed topically due to their hydrophobic properties. CBD skin absorption is greater than that for THC.
    The formulation of cannabinoids in various oils may facilitate absorption, but adequate data is lacking on the ideal composition or vehicle.

Pharmacodynamics of Cannabis

- There are two known cannabinoid receptors: [3, 4]
  - Cannabinoid Brain receptor -1 (CB_1).

This receptor is present throughout the central nervous system (CNS) with high concentrations in the neocortex, hypothalamus, hippocampus and cerebellum.

They are also found peripherally on axon terminals in the liver, lungs, pancreas, skeletal muscle and fat cells.

The receptor is a G-protein coupled receptor.
- Inhibitory modulation of many neurotransmitters including norepinephrine, dopamine, serotonin, GABA, and acetylcholine.
- Cannabinoid Brain receptor -2 (CB_2)

  This receptor is found primarily on immune cells.
- The two major cannabinoids are tetrahydrocannabinol (THC) and Cannabidiol (CBD).
  - THC is the psychoactive constituent of cannabis and is responsible for its intoxicating effects. It is mainly active on the CB_1 receptor.
  - Cannabidiol does not appear to have the psychoactive effects of THC.

Emergency Department (ED) Presentations

- There has been a significant increase in patient visits to the ED after both acute and chronic cannabis use. Part of the reason is the higher potency of THC as well as increased recreational use of cannabis.
- There are several unique presentations that the clinician must be aware of when faced with a patient who uses cannabis:
  - Acute Cannabis Intoxication.
  - Intoxication secondary to Synthetic Cannabinoid Receptor Agonists.
  - Cannabis Use Disorder.
  - Cannabis Withdrawal.
  - Cannabinoid Hyperemesis Syndrome.
  - Accidental Pediatric Ingestions.

Acute Cannabis Intoxication [2, 5]

- The clinical presentation of acute cannabis intoxication varies with:
  - The amount of cannabis consumed.
  - The route of consumption.

    Increased toxicity is seen with inhaled cannabis given the quick onset and higher bioavailability.

    In contrast, prolonged symptoms are seen with the oral route.
  - Concentrations of THC in the compound or resin.

    There is a trend toward increasing THC concentration seen in available products when compared to two decades ago.
  - Drug formulation.
  - Age of the patient.

The concentration of cannabinoids available at receptors will vary with the individual's pharmacokinetic and pharmacodynamic response to cannabis.

- Clinical Signs and symptoms of Acute Cannabis Intoxication include:
  - Tachycardia.
  - Hypertension or orthostatic hypotension.
  - Conjunctival injection.
  - Dry mouth.
  - Increased appetite.
  - Hallucinations, depersonalization, paranoia and other features of psychosis may occur especially with higher concentrations of THC.
  - Cognitive impairment may occur including issues with short term memory and concentration.
  - Psychomotor changes including impaired coordination and deceased reaction time.
  - Children more frequently present after oral ingestion of cannabis products with CNS depression whereas adults more frequently present with CNS excitation.
- Management of acute toxicity
  - Most patients are treated with general supportive measures.
  - For patients that have severe agitation, benzodiazepines are the treatment of choice.
  - Patients who present with dehydration can be treated with intravenous fluids.

Intoxication Secondary to Synthetic Cannabinoid Receptor Agonists (SCRAs) [6–8]

- Most of the synthetic cannabinoids are developed for illegal recreational use.
  - They are known by various street names: K-2, Black mamba, crazy clown, spice, Spice2, Smoke, and Summit.
- These compounds have distinct chemical and pharmacodynamic properties compared to plant-derived cannabis products.
- Due to their higher THC concentration (compared to cannabis plant), and strong binding affinity to the endocannabinoid receptors (that are active at the serotonin and NDMA receptors) SCRAs have been associated with severe toxicity.
  - Many of the synthetic cannabinoids were developed with the aim of increasing euphoria. As a result, their effects on the endocannabinoid receptors are much greater than THC.
  - JWH-018 has 4 times the affinity for the CB_1 receptors compared to THC, and 20 times the affinity for CB_2 receptors.
- In addition to the psychoactive effects associated with cannabis use, SCRA ingestion also produces a sympathomimetic toxidrome, which can result in:
  - Diaphoresis.

- Agitation.
- Restlessness.
- Seizures.
- These compounds are also the preferred agent of choice by users who require drug monitoring because these compounds are not detectible on standard drug screens.
- SCRA products have been associated with outbreaks of severe sympathomimetic toxicity and as such are a significant public health concern.
- Management of SCRA Intoxication is mainly supportive and includes:
  - Adequate hydration (IV fluids).
  - Managing agitation with benzodiazepines as first line agents.
    - Consider additional agents such as haloperidol or ketamine if escalating therapy is required.
  - Treatment of seizures with benzodiazepines.
  - The observation period should be prolonged for up to 24 hours or more if there is a high index of suspicion for SCRA toxicity.

Cannabis Use Disorder (Problematic Marijuana Users)

- Individuals who demonstrate compulsive patterns of use or who experience harmful consequences secondary to repeated use of cannabis can be diagnosed with Cannabis Use Disorder.
- The DSM V criteria for this diagnosis involve the above problematic patterns of use and two or more of the following within a 12-month period: [9]
  - Increasing consumption over time.
  - Difficulty moderating use.
  - Cravings that are sustained.
  - Recurrent use leading to significant failure at school, work or home.
  - Continued use despite social or interpersonal problems exacerbated by cannabis use.
  - Important activities abandoned because of cannabis use.
  - Use in situations that result in physical danger.
  - Continued use despite known health consequences.
  - Tolerance.
  - Withdrawal.
- Severity is graded using the following criteria:
  - Mild – 2–3 of the criteria are met.
  - Moderate – 4–5 criteria.
  - Severe – 6 or more criteria.
- These patients are not routinely managed in the Emergency Department. However, they should be referred to an addiction specialist or service for appropriate care.

Cannabis Withdrawal

- Individuals who have used cannabis chronically and stop abruptly can experience symptoms of withdrawal.
- The DSM-5 identifies the symptoms of Cannabis Withdrawal as having the following up to a week after stopping use: [9]
    - Anxiety and restlessness.
    - Depression or irritability.
    - Insomnia, as well as odd dreams.
    - Physical symptoms, such as tremors or headache.
    - Decreased appetite.
- Management of cannabis withdrawal is mainly supportive care.
    - Although cannabis withdrawal is uncomfortable and unpleasant for the patient, it does not cause seizures and is not life-threatening.

Cannabinoid Hyperemesis Syndrome (CHS) [10–14]

- CHS presents as intractable nausea, vomiting and abdominal pain that is associated with chronic regular cannabis use.
- Patients will describe a history of frequent or even compulsive hot showers or baths for temporary relief of symptoms. This is pathognomonic for CHS.
- The syndrome was first described in 2004. Since then, there has been increased recognition of the condition and several therapies have been shown to be effective.
- CHS Accounts for up to 6% of Emergency Department presentations with recurrent vomiting.
- Patients with CHS are typically younger, male gender and have a history of frequent and heavy cannabis use.
    - Most patients with CHS report using cannabis daily.
- These patients often have recurrent presentations to the Emergency Department (ED) for these symptoms and undergo a number of diagnostic procedures including CT scans, Ultrasound and endoscopy.
- One study described an average of 7 ED visits before the diagnosis was made.
- Three phases of the syndrome have been described:
    - Prodromal or pre-emetic phase which can last for months to years.
        - During this phase, patients often experience severe early morning nausea, vomiting and abdominal discomfort.
        - They usually maintain normal eating patterns and monitor their weight.
        - Paradoxically, they may actually increase their use of cannabis in an effort to treat their symptoms.
    - The Hyperemesis phase is characterized by periods of severe and intense nausea and profuse vomiting, associated with diffuse abdominal pain.
        - These patients describe their vomiting as incapacitating and can occur frequently and without warning.

Many patients will present with dehydration and weight loss.

Patients will often describe transient relief with hot baths or showers.

It is during this phase that patients tend to present to the emergency department.

- Recovery phase can last for days to months and corresponds with the cessation of cannabis.

During this phase, patients manifest normal eating and bathing patterns and often gain back the weight that was lost.

However, symptoms can recur if the patient again uses cannabis.

Pathophysiology of CHS [15]

- The mechanism of CHS is not well understood. CHS may be due to dysregulation of the endocannabinoid system.
  - CB_1 and CB_2 receptors are not only found in the brain but throughout the gastrointestinal tract.
  - The endocannabinoid system is thought to play a role in gastrointestinal motility, appetite, and nausea/vomiting.
- However, the pattern of symptom relief with heat exposure in the form of hot baths and showers suggests a possible functional relationship between cannabinoid receptors and Transient Receptor Potential Vanilloid-1 ($TRPV_1$) receptors, which are activated by heat and are found in close proximity or co-located with endocannabinoid receptors.
  - Low-level stimulation of the TRPV1 receptors may result in nausea and vomiting.
  - Many cannabinoids are TPRV1 agonists.
  - Repeated stimulation of theses receptors by these cannabinoids can induce persistent nausea and vomiting.
  - Overstimulation of TPRV1 receptors by heat can lead to amelioration of the symptoms.

Diagnosis of CHS [10, 16]

- CHS has usually been a diagnosis of exclusion in the context of a history of chronic regular cannabis use. However, with better recognition of this condition, clinicians are getting better at diagnosing it earlier.
- CHS has often been confused with Cyclic Vomiting Syndrome given the similar presentation and lack of efficacy with traditional anti-emetic medications.
- These patients will often have had many tests (laboratory, imaging, endoscopy) that have been non-diagnostic.
- Characteristics that are helpful in the diagnosis include:
  - Patient characteristics.
    Male gender.

- Cannabis use.
    Daily, heavy or long term use.
- Presenting symptoms.
    Severe nausea and vomiting.
    Abdominal pain.
- Alleviating factors.
    Hot showers or baths are helpful.
    Symptoms resolve when the patient is not using cannabis.

Management [17–19]

- Symptoms are often refractory to the usual care for nausea and vomiting.
- First line anti-emetic treatments such as dimenhydrinate, metoclopramide, and ondansetron should be tried, but often do not effectively resolve patient symptoms.
- Antipsychotic medications such as haloperidol and benzodiazepines have been used with some effect.
- Topical capsaicin applied to the abdomen and dorsal aspect of arms may provide relief of symptoms [20, 21].
    - Capsaicin, an active component of chili peppers, like high temperatures, activates $TRPV_1$ receptors, which may explain its potential for providing relief of symptoms.
    - Suggested dose of capsaicin ointment is 0.075% topically to abdomen or back of arms three times daily in a layer approximately 1 mm thick across the chosen areas. It is suggested that gloves be worn during application to avoid burning sensation to the hands.
- The only way to prevent recurrence is cessation of cannabis use.

Accidental Pediatric Ingestions [2, 5, 22–24]

- Pediatric poisonings occur most often occur through inadvertent ingestions of edibles that resemble normal food (examples include brownies or gummy bears), although ingestions of marijuana in other forms, such as a cannabis filled rolls (known as blunts), cigarettes, or "joints", that are accessible to children can occur as well.
- Toddlers have the highest rate of unintentional ingestions [22].
- Rates of accidental pediatric exposures to cannabis have been shown to increase with the legalization of marijuana.
- Lethargy is the most common presenting sign in cases of unintentional pediatric ingestion.
    - Other signs may include ataxia, mydriasis, tachycardia and hypotonia.
- In severe cases seizures, coma and respiratory depression have been reported.
    - Intubation for airway protection should be considered in comatose children.

- – Neuroimaging is often required to rule out a structural cause for the altered level of consciousness.
- Consider other toxicological or organic causes in children that present with an altered level of consciousness or coma.
- Perform a urine drug screen in the comatose pediatric patient with normal neuroimaging.
  - – This test may be essential for diagnosis and to aid Child Protective Services.
- Involve Child Protective Services in all cases of accidental pediatric cannabis exposure.
- Educate adult patients who present with cannabis related toxicity on the importance of childproof storage of cannabis product.

Medical Conditions Associated with Cannabis Use

- Given the relatively recency of cannabis legalization, robust evidence linking cannabis use to other medical conditions is lacking. However, there have been observational studies and case reports that suggest a correlation between cannabis use and:
  - – Cardiovascular Risk. [25–28]
      Use of cannabis has been correlated to increased risk of myocardial infarction, stroke, and atrial fibrillation.
      The postulated mechanism includes:
      - Increase myocardial oxidative stress.
      - Hyperadrenergic state with resultant myocardial demand and ongoing stress to vasculature.
      - Myocardial depressant.
      - Procoagulant activity.
  - – Mental Health Risk. [4, 29–31]
      It is postulated that cannabis use under age 25, while the brain is still developing, is associated with increased risk of:
      Schizophrenia.
      Bipolar Disorder.
      Anxiety Disorders.
      Depression.
  - – Pulmonary Risk. [32]
      Evidence that regular cannabis use contributes to the development or worsening of obstructive lung diseases such as COPD and asthma is conflicting.

Key Concepts

- Consider Cannabis Hyperemesis Syndrome in the differential diagnosis for patients with recurrent nausea, vomiting and abdominal pain, and a normal work-up.

- Cannabis intoxication presents differently depending on formulation, amount, strength and route of consumption and varies with age.
- Consider cannabis exposure in the pediatric patient with altered level of consciousness and other neurologic signs and symptoms.

Laws and Regulations Referenced
Cannabis Act, SC 2018, c 16.

References

1. Hasin DS. US epidemiology of cannabis use and associated problems. Neuropsychopharmacology. 2018;43(1):195–212.
2. Blohm E, Sell P, Neavyn M. Cannabinoid toxicity in pediatrics. Curr Opin Pediatr. 2019;31(2):256–61.
3. Mackie K. Cannabinoid receptors: where they are and what they do. J Neuroendocrinol. 2008;20(Suppl 1):10–4.
4. Sachs J, McGlade E, Yurgelun-Todd D. Safety and toxicology of cannabinoids. Neurotherapeutics. 2015;12(4):735–46.
5. Noble MJ, Hedberg K, Hendrickson RG. Acute cannabis toxicity. Clin Toxicol (Phila). 2019;57(8):735–42.
6. Cohen J, et al. Clinical presentation of intoxication due to synthetic cannabinoids. Pediatrics. 2012;129(4):e1064–7.
7. Adams AJ, et al. "zombie" outbreak caused by the synthetic cannabinoid AMB-FUBINACA in New York. N Engl J Med. 2017;376(3):235–42.
8. Kasper AM, et al. Severe illness associated with reported use of synthetic cannabinoids— Mississippi, April 2015. MMWR Morb Mortal Wkly Rep. 2015;64(39):1121–2.
9. American Psychiatric Association. Diagnostic and statistical manual of mental disorders. 5th ed. Arlington, VA; 2013.
10. Lapoint J, et al. Cannabinoid hyperemesis syndrome: public health implications and a novel model treatment guideline. West J Emerg Med. 2018;19(2):380–6.
11. Chocron Y, Zuber JP, Vaucher J. Cannabinoid hyperemesis syndrome. Bmj. 2019;366:l4336.
12. Allen JH, et al. Cannabinoid hyperemesis: cyclical hyperemesis in association with chronic cannabis abuse. Gut. 2004;53(11):1566–70.
13. Hernandez JM, Paty J, Price IM. Cannabinoid hyperemesis syndrome presentation to the emergency department: a two-year multicentre retrospective chart review in a major urban area. CJEM. 2018;20(4):550–5.
14. Khattar N, Routsolias JC. Emergency department treatment of cannabinoid hyperemesis syndrome: a review. Am J Ther. 2018;25(3):e357–61.
15. Perisetti A, et al. Cannabis hyperemesis syndrome: an update on the pathophysiology and management. Ann Gastroenterol. 2020;33(6):571–8.
16. Sorensen CJ, et al. Cannabinoid hyperemesis syndrome: diagnosis, pathophysiology, and treatment-a systematic review. J Med Toxicol. 2017;13(1):71–87.
17. Reinert JP, et al. Management of pediatric cannabinoid hyperemesis syndrome: a review. J Pediatr Pharmacol Ther. 2021;26(4):339–45.
18. Richards JR. Cannabinoid hyperemesis syndrome: pathophysiology and treatment in the emergency department. J Emerg Med. 2018;54(3):354–63.
19. Zhu JW, et al. Diagnosis and acute management of adolescent cannabinoid hyperemesis syndrome: a systematic review. J Adolesc Health. 2021;68(2):246–54.

20. Graham J, Barberio M, Wang GS. Capsaicin cream for treatment of cannabinoid hyperemesis syndrome in adolescents: a case series. Pediatrics. 2017;140:6.
21. McConachie SM, et al. Efficacy of capsaicin for the treatment of cannabinoid hyperemesis syndrome: a systematic review. Ann Pharmacother. 2019;53(11):1145–52.
22. Wang GS, et al. Association of unintentional pediatric exposures with decriminalization of marijuana in the United States. Ann Emerg Med. 2014;63(6):684–9.
23. Richards JR, Smith NE, Moulin AK. Unintentional cannabis ingestion in children: a systematic review. J Pediatr. 2017;190:142–52.
24. Wang GS, et al. Unintentional pediatric exposures to marijuana in Colorado, 2009-2015. JAMA Pediatr. 2016;170(9):e160971.
25. Singh A, et al. Cardiovascular complications of marijuana and related substances: a review. Cardiol Ther. 2018;7(1):45–59.
26. Mittleman MA, et al. Triggering myocardial infarction by marijuana. Circulation. 2001;103(23):2805–9.
27. Patel RS, et al. Marijuana use and acute myocardial infarction: a systematic review of published cases in the literature. Trends Cardiovasc Med. 2020;30(5):298–307.
28. Rumalla K, Reddy AY, Mittal MK. Recreational marijuana use and acute ischemic stroke: a population-based analysis of hospitalized patients in the United States. J Neurol Sci. 2016;364:191–6.
29. Gibbs M, et al. Cannabis use and mania symptoms: a systematic review and meta-analysis. J Affect Disord. 2015;171:39–47.
30. Degenhardt L, et al. The persistence of the association between adolescent cannabis use and common mental disorders into young adulthood. Addiction. 2013;108(1):124–33.
31. Mustonen A, et al. Adolescent cannabis use, baseline prodromal symptoms and the risk of psychosis. Br J Psychiatry. 2018;212(4):227–33.
32. Ghasemiesfe M, et al. Marijuana use, respiratory symptoms, and pulmonary function: a systematic review and meta-analysis. Ann Intern Med. 2018;169(2):106–15.

Part III

Cannabis Use in Specific Populations

Occupational Considerations Related to Cannabis Use

Gregory L. Smith

Introduction

In 2017 15% of a nationally representative sample of 16,280 US adults used cannabis in the past year, and the rate was 20% in those states with recreational cannabis laws [1]. Cannabis for treating medical conditions is currently recommended by physicians to over four million adults and the number is increasing rapidly in jurisdictions that have medical cannabis regulations. In addition, 7% of the adult population in the US uses Cannabidiol (CBD) on a regular basis and this number is expected to increase to 10% by 2025 [2]. This rapid and exponential increase in cannabinoid use has raised safety concerns in the workplace. Cannabis has known dose-dependent effects on cognition, psychomotor activities, balance and coordination. It can have adverse effects of anxiety, panic, and psychosis. In addition, alcohol and some other prescription drugs are synergistic with the effects of cannabis.

In the US, decades old employer and federal urine drug testing (UDT) policies and programs are undergoing repeated legal challenges due to the state legal, but federally illegal use of medical cannabis by employees that are protected by the state and federal American with disabilities Acts (ADA). Significant changes to employer policies on the use of cannabis are forthcoming.

Medical Cannabis in the Workplace

- Employers in the US are currently in an ever-increasing battle between complying with federal drug testing requirements for safety sensitive jobs, and risking lawsuits under disability and discrimination lawsuits.

G. L. Smith (✉)
Cannabinoid Research, Nex-Therapeutics LLC, St. Pete Beach, FL, USA
e-mail: DrSmith@Nex-Therapeutics.com

© Springer Nature Switzerland AG 2022
R. Valani (ed.), *Cannabis Use in Medicine*,
https://doi.org/10.1007/978-3-031-12722-9_13

- In the United States, State and Federal American's with Disability Act (*Americans With Disabilities Act* of 1990, Pub. L. No. 101-336, 104 Stat. 328 (1990)) protect the rights of those with a disability. Discrimination [3, 4].
  - In Canada, the Accessible Canada Act (Accessible Canada Act, SC 2019, c 10) aims to remove and prevent barriers for persons with disabilities by 2040 [5]. Each province also has additional laws to help develop standards. For example, Ontario has the *Accessibility for Ontarians with Disabilities Act* (Accessibility for Ontarians With Disabilities Act, 2005, SO 2005, c 11) which is designed to reduce discrimination for persons with disabilities by 2025 [6].
- The US federal government has remained steadfast in classifying cannabis as a Schedule I drug.
  - Schedule I drugs, under the federal *Controlled Substances Act* (CSA), are those for which the *Drug Enforcement Administration* (DEA) determines there is a high potential for abuse and for which there is no currently accepted medical use.
  - The DEA reviewed its classification of cannabis in 2016 and kept it on the Schedule I list.
  - Schedule I drugs cannot be prescribed by doctors or distributed at pharmacies. Possession and distribution of a Schedule I substance can be criminally prosecuted in federal court. Federal requirements for drug-free workplaces still require that employees test negative for THC (Δ9-Tetrahydrocannabinol) along with several other illegal drugs [7].
- States that have legalized cannabis have a variety of ways they treat medical cannabis use by employees. (See Chap. 2 on the legal aspects of cannabis)
  - Some protect an employer's right to maintain a cannabis-free workplace; others make it difficult for employers to regulate cannabis use by employees.
  - This is also complicated by state court rulings, which sometimes add additional protections for cannabis-using employees.
  - Workplace protections are, for the most part, limited to medical cannabis, with almost no protections for employees who use cannabis recreationally. However, in the 10 states and Washington, D.C., where cannabis is recreationally legal, the laws are still not cut and dry [7].
- Regular cannabis use, whether it be used for recreational or medical purposes, can result in positive UDT for THC as much as 30 days after the last use. See Chap. 5 on analytical testing of cannabis.
  - THC and other highly fat-soluble compounds have a very long half-life of elimination and can be detected in urine up to weeks after last use among heavy users [8]. This has highlighted the issues of THC use while off-duty or on vacation, but not while no the job [9].
- In general states have taken the policy of prohibiting employers from disciplining an employee for a positive THC test alone if the employee is a certified medical cannabis patient [10].
  - Essentially, a positive urine drug test for THC in and of itself is not grounds for adverse action. In those legal states where the medical cannabis legislation

doesn't specifically address employer discipline, state anti-discrimination laws requiring 'reasonable accommodation' have been used.

CBD Use in the Workplace

- CBD is available over the counter in all 50 states of the USA, and is being used for dozens of symptoms and conditions. This legal nutraceutical is made from hemp plants that usually has less than 0.3% THC [11].
  - Therefore, products made using certified good manufacturing practices (cGMP) should not contain enough THC to cause a UDT to turn positive for THC metabolites.
- CBD is also available by prescription for intractable childhood seizures. See Chap. 9 on neurological diseases and cannabinoid treatment.
  - The use of CBD should not cause any concern for educated employers because it has no psychoactive effects, such as euphoria, and has no known impairing effects on cognition, coordination or balance.
  - Unfortunately, there are some lower quality CBD products that can contain many times more than the 0.3% THC on the label [12]. These 'hot' CBD formulations may cause impairment from the higher THC content and positive UDT for THC.

Functional Impairment

- Occupational functional impairment refers to deficits in cognitive, psychomotor, physical, behavioral functioning or any combination of these such that the work activity is adversely impacted, or a safety hazard is created.
  - The endpoint of this functional impairment would be an accident, injury or death to the impaired worker, or others.
  - Historically, much of the research related to occupational functional impairment involves alcohol impairment.
- Driving and certain work activities are recognized to be a highly complex activity involving a wide range of cognitive, perceptual, and motor activities that take place in a complex, dynamic environment.
  - Decades of research on driving and work activities have identified various tasks involved in the performance of these activities.
  - The most critical tasks are determined empirically through crash and accident data analyses, or through experimental studies.
  - The following five domains have been determined by expert panels to be relevant to driving and work abilities: [13, 14]

        A need for alertness/ arousal during the process.
        Having the attention and processing speed to understand all the stimuli.
        Ensuring adequate reaction time and psychomotor functions to correct any course of action.

Intact sensory-perceptual functions.

High level executive functions.

- An intoxicant that impairs performance in any of these domains at a magnitude known to be associated with increased crash or accident risk is presumed to have a negative impact on safety.
- THC, especially in higher doses, can also increase anxiety and cause panic attacks, and in some cases cause paranoia and hallucinations. These effects have been noted to last as long as 24 h [14].
- A meta-analysis of 60 studies concluded that cannabis causes impairment in every performance area that can reasonably be connected with safe operation of a vehicle or safe work practices, such as tracking, motor coordination, visual functions, and particularly complex tasks that require divided attention [15].

  Another study showed that attentiveness, vigilance, perception of time and speed, and use of acquired knowledge are all affected by cannabis [16].

  A major factor to consider with these studies is the concentration of THC being used, the amount being consumed, and other confounding variables such as use of other medications / substances that can also cause impairment.

Work Performance

- It is difficult to determine if THC has causes impairment in a dose-dependent manner due to the heterogeneity of studies, route of administration, and individual metabolism.
- When it comes to cannabis, impairment of psychomotor or cognitive abilities is impacted primarily by the amount of THC that is absorbed, and the rate at which it is absorbed.
  - THC and the other cannabinoids are fat soluble (highly lipophilic). Soon after entering the blood stream, over 90% of the cannabinoids are absorb into the body's fat stores, and then slowly redistributed back into the circulation over several days.
- THC that is absorbed via the lungs (smoking and vaping) is absorbed directly into the blood circulation within minutes. There is a rapid peak plasma level between 9-23 min followed by a gradual sloping off of the plasma concentration especially over 90 min [17]. See Chap. 3 on Pharmacology of cannabis.
  - Ingested THC is absorbed much more slowly and variably over two to four hours and goes through the first-pass effect of the liver, where 90% of it is transformed into the metabolite, 11-OH-THC that is at least 1.4 times as potent on affecting psychomotor and cognitive abilities as THC [17, 18].
- Cannabis users tend to overestimate their impairment, and consequently employ compensatory strategies.
  - Cannabis users perceive their driving under the influence as impaired and more cautious [19].

- Detrimental effects of cannabis are more pronounced with highly automatic driving functions than with more complex tasks that require conscious control, whereas with alcohol produces an opposite pattern of impairment.
- Driving and simulator studies show that detrimental effects vary in a dose-related fashion, and are more pronounced with highly automatic driving functions, but more complex tasks that require conscious control are less affected. Lower levels of THC, when combined with alcohol, are sufficient to cause obvious impairment [20]. See Fig. 13.1.
- Experimental studies have shown that functional impairment, reaches a maximum in minutes to hours, depending on inhalation or ingestion, then rapidly decreases over 2 ½ hours for inhalation and up to 6 h for ingestion.
  - This makes it much harder to generate blood level versus impairment curves for cannabis than it is for alcohol [22].
- Because of these difficulties, epidemiological studies have also shown inconsistent effects, some finding decreased or no risk from driving while smoking cannabis, and others increased risk [23, 24].
  - Most studies are fraught with methodological problems that could lead to underreporting of drug use or misclassification of experimental subjects into or out of the cannabis-using category, confounding results.

Figure 13.1 above shows that as the serum level of THC increased above 5 ng/ml there is an exponential increase in the odds ratio of the risk of an accident. Other

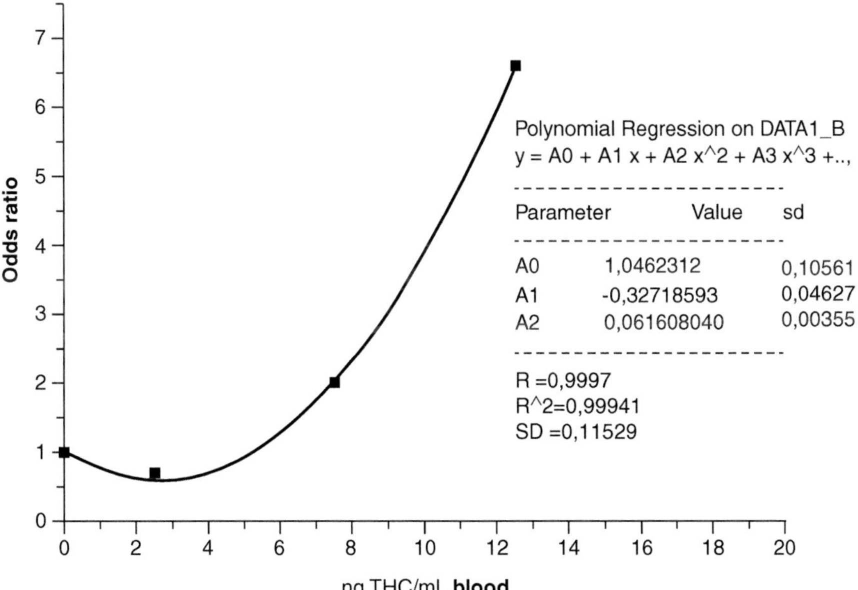

Fig. 13.1 (with permission). Correlation between THC concentration in whole blood and accident risk (from Grotenhermen et al. (2007) based on data from Drummer et al.) [21]

studies have suggested that serum levels below 7 ng/ml are not associated with elevated accident risk, with levels of 7-10 ng/ml equating with blood alcohol concentrations of 0.05% [21].

Specific Cognitive, Psychomotor and Physical Functional Impairment

- The most consistent and validated evidence for THC effects are with short-term memory, focus, selective attention, balance, divided, attention and sustained attention [25].
- Alcohol, on the other hand, has been shown to impair:
  - Critical flicker fusion (the frequency at which flickering lights are perceived as continuous).
  - Short-term memory.
  - Pursuit tracking.
  - Divided attention.
  - Signal detection.
  - Hazard perception.
  - Reaction time.
      Alcohol, more than THC, affects reaction time [26].
  - Attention.
  - Concentration.
  - Hand-eye coordination.

Testing for Impairing Drugs

- Alcohol and most psychoactive drugs (licit and illicit) are water-soluble.
  - Water-soluble drugs are easier to measure from available testing of breath, saliva, urine, or blood.
  - There are decades of robust research data on the correlation of plasma levels of water-soluble intoxicants with the level of functional impairment.
- Alcohol and several drugs cause functional impairment by anesthetizing certain brain centers in well described dose-response curves.
  - Other drugs, such as benzodiazepines and opioids are diffuse CNS depressants and cause functional impairment through this mechanism [26].
- THC causes functional impairment by stimulating CB_1 receptors of certain centers in the brain that relate to balance, coordination, attention, and memory.
  - Frequent users of THC will tend to have down-regulation of their CB_1 receptors, meaning that there are fewer CB_1 receptors on neuronal cell membranes and binding is less efficient [27].
  - This means that at a specific plasma level of THC there will be much less functional impairment than with a person with normal numbers of CB_1 receptors.

- A small percentage of persons have genetic variation in the binding efficiency of the CB_1 receptors, resulting in more or less functional impairment in response to specific plasma levels of THC [28]. See Chap. 6 on the role of genetics in the use of cannabis.
- The well known 'entourage effect' of CBD, minor cannabinoids and terpenes present in cannabis, results in direct and indirect synergism of the THC at the CB_1 receptor via allosteric and similar mechanisms [29].
    - This means that a specific plasma level of THC can have widely variable functional impairment depending on the presence of these 'entourage' components.
- THC that is ingested goes through the first-pass effect in the liver. About 90% of the THC is immediately metabolized to 11-OH-THC meaning that the plasma level of THC is very low after ingestion as it has been metabolized by the time it enters the blood stream [18].
    - More importantly, 11-OH-THC is at least 1.4 times as potent as THC at the CB_1 receptor, resulting in higher levels of impairment compared to a similar plasma level of THC.
- The net effect of these mechanisms is that there are no generalizable dose-response curves for THC as there are with most other intoxicants.
    - Using biologic specimens to determine functional impairment continues to be unfruitful for THC. This has made it essentially impossible to have a specific cut-off plasma level to determine impairment from THC. A positive result does not document impairment, or even recent use [30].
- Some technology doesn't attempt to measure functional impairment from THC level.
    - Instead, the test is to determine if there has been very recent use of THC.
        The assumption being that 'very recent use' within past 2 h, is tantamount to the THC having an 'influence' on functioning, a legal silhouette of functional impairment.
        This testing includes plasma or saliva levels of THC and metabolites of THC. By using certain ratios of THC to metabolites the test purports to confirm recent use of THC [31, 32].
        The wide variation in absorption of fat-soluble cannabinoids, the difference in THC metabolites from various routes of administration, and the very long half-life of cannabinoids makes this testing methodology untenable.

Synergistic Effects of THC, Alcohol and Other Prescription Drugs

- Studies have shown that opioids, benzodiazepines, antihistamines and alcohol are all synergistic with the impairing effects of THC. Specifically, cannabis and alcohol, when used together, have additive or even multiplicative effects on impairment [24].

- In essence this means that using THC in conjunction with alcohol or some other drugs could result in an unexpected, unintentional much higher level of impairment, from that of the THC alone.
  - A study of motor vehicle deaths calculated an Odds Ratio of 0.7 for cannabis use alone, 7.4 for alcohol use, and 8.4 for cannabis and alcohol use combined [33].
- Cannabis and alcohol acutely impair several driving and work-related skills in a dose-related fashion.
  - For reasons discussed above, the effects of cannabis vary more between individuals than they do with alcohol because of tolerance, differences in smoking technique, different absorptions of THC, receptor down-regulation and genetic variation in CB_1 receptors.
  - The risk from driving under the influence of both alcohol and cannabis is greater than the risk of driving under the influence of either alone [23].
- Recreational cannabis is often compared to alcohol, but for employers, there is a major difference.
  - Currently it is nearly impossible to assess a cannabis user's level of impairment.

    A simple and non-invasive breath or saliva test can tell an employer on the spot how impaired an employee is due to alcohol use and can allow a timely decision to take an employee out of a dangerous position.
  - While science can tell us that a blood alcohol concentration of .08 percent has specific effects on a person's functioning, science cannot tell us what effect a certain concentration of THC will have on an individual.

    However, according to Fig. 13.1, levels of THC above 5 ng/ml show increasing dose-response risk of an accident.
- A new wearable technology device is currently undergoing evaluation for use in the Occupational Health setting and for roadside use to detect physical and cognitive impairment from cannabis. (personal communication – Functional Mimetics Technologies "HiBit®".)

Urine Drug Testing—State Vs. Federal law

- There are many safety sensitive job categories and classifications including commercial drivers, pilots, and other jobs that have federally mandated UDT.
  - THC is by the far the most common "positive" of the "SAMHSA-5" panel drug test. This panel of five drugs (THC, cocaine, amphetamines, opiates and phencyclidine (PCP)) was developed in the 1980s under the Drug-Free Workplace Act. There are also federally mandated breath alcohol test (BAT) requirements [8].
- It is interesting that two FDA approved drugs that are legally prescribed and made from dronabinol (Marinol®, and Syndros®) will cause positive UDTs for THC. Dronabinol is a THC analog and is notorious impairing with significant reports of euphoria [34].

CBD and Positive Urine Drug Tests (UDTs)

- As was mentioned above, poor-quality CBD formulations may be 'hot' meaning they have higher levels of THC in them then is written on the label.
- Use of these over-the-counter products could cause of positive UDT for THC. However, if the CBD is being used for a protected medical condition, this may not support adverse employer actions.

Future

- The state of New Jersey in July 2019, amended the Medical Marijuana legislation to include protections for employees and employers.
  - Employers are now prohibited from taking any "adverse employment action" against workers who are registered qualifying patients.
  - Employees and job applicants are also given the right to explain positive drug test results.
  - At the same time the amendments stated that businesses are also not required to commit to any policy that would cause them to be in violation of federal law or that would result in the loss of a federal contract or funding.
- The trend is for more states and jurisdictions, forced by judicial rulings, to provide clearer and greater protection to employees that use medical cannabis.
- Perhaps we can learn the most about what the future portends from Canada, where cannabis has been federally legalized for recreational use.
  - The Canadian Public Services Health and Safety Association future direction is toward preventing the impairment caused by THC, as opposed to looking for evidence of THC use. This will require better employee and management training for warning signs of impairment and better means of identifying impaired from unimpaired workers and drivers.

References

1. Keyhani S, et al. Risks and benefits of marijuana use: a National Survey of U.S. Adults. Ann Intern Med. 2018;169(5):282–90.
2. Owram K. *Much-hyped US CBD Market $16 Billion by 2025*. 2019.
3. Pirone J. *Marijuana in the workplace*. 2019.; https://ohsonline.com/articles/2019/03/01/marijuana-in-the-workplace.aspx.
4. Americans With Disabilities Act of 1990, Pub. L. No. 101-336, 104 Stat. 328 (1990).
5. Accessible Canada Act, SC 2019, c 10.
6. Accessibility for Ontarians With Disabilities Act, 2005, SO 2005, c 11.
7. Code UF. *41 U.S. Code § 8102—Drug-free workplace requirements for Federal contractors | U.S. Code | US Law | LII / Legal Information Institute*, U.F. Government, Editor. 1988.
8. Hadland SE, Levy S. Objective testing: urine and other drug tests. Child Adolesc Psychiatr Clin N Am. 2016;25(3):549–65.
9. Render, H. *Firing employee for using medical marijuana off duty is not illegal*. 2015.; https://www.lexology.com/library/detail.aspx?g=71229af9-6f10-4d59-9916-cfee656f3d74.

10. Phillips JA, et al. Marijuana in the workplace: guidance for occupational health professionals and employers: joint guidance statement of the American Association of Occupational Health Nurses and the American College of Occupational and Environmental Medicine. Workplace Health Saf. 2015;63(4):139–64.

11. Agriculture, N.I.o.F.a., *Industrial Hemp | National Institute of Food and Agriculture*, USDA, Editor. 2019, US Department of Agriculture.

12. Bonn-Miller MO, et al. Labeling accuracy of cannabidiol extracts sold online. JAMA. 2017;318(17):1708–9.

13. Diamond A. Executive functions. Handb Clin Neurol. 2020;173:225–40.

14. Transportation, U.D.o. *Driving expert panel report: a concensus protocol for assessing the potential of drugs to impair driving.* 2011, NHTSA.

15. Berghaus G, G.B. *Medicine and Driver Fitness—findings from a metaanalysis of experimental studies as basic information to patients, physicians and experts.* in *Proceedings of the 13th International Conference on Alcohol, Drugs and Traffics Safety.* 1995. *Adelaide, Australia.*

16. Liguori A, Gatto CP, Robinson JH. Effects of marijuana on equilibrium, psychomotor performance, and simulated driving. Behav Pharmacol. 1998;9(7):599–609.

17. Huestis MA. Human cannabinoid pharmacokinetics. Chem Biodivers. 2007;4(8):1770–804.

18. Huestis MA, Henningfield JE, Cone EJ. Blood cannabinoids. I. Absorption of THC and formation of 11-OH-THC and THCCOOH during and after smoking marijuana. J Anal Toxicol. 1992;16(5):276–82.

19. Macdonald S, et al. Testing for cannabis in the work-place: a review of the evidence. Addiction. 2010;105(3):408–16.

20. Jones AW, Holmgren A, Kugelberg FC. Driving under the influence of cannabis: a 10-year study of age and gender differences in the concentrations of tetrahydrocannabinol in blood. Addiction. 2008;103(3):452–61.

21. Grotenhermen F, et al. Developing limits for driving under cannabis. Addiction. 2007;102(12):1910–7.

22. Macdonald S, et al. Injury risk associated with cannabis and cocaine use. Drug Alcohol Depend. 2003;72(2):99–115.

23. Sewell RA, Poling J, Sofuoglu M. The effect of cannabis compared with alcohol on driving. Am J Addict. 2009;18(3):185–93.

24. Hartman RL, Huestis MA. Cannabis effects on driving skills. Clin Chem. 2013;59(3):478–92.

25. Crean RD, Crane NA, Mason BJ. An evidence based review of acute and long-term effects of cannabis use on executive cognitive functions. J Addict Med. 2011;5(1):1–8.

26. Zoethout RW, et al. Functional biomarkers for the acute effects of alcohol on the central nervous system in healthy volunteers. Br J Clin Pharmacol. 2011;71(3):331–50.

27. Bolognini D. *Pharmacological properties of the phytocannabinoids Δ9-tetrahydrocannabivarin and cannabidiol.* 2010.

28. Stout SM, Cimino NM. Exogenous cannabinoids as substrates, inhibitors, and inducers of human drug metabolizing enzymes: a systematic review. Drug Metab Rev. 2014;46(1):86–95.

29. Laprairie RB, et al. Cannabidiol is a negative allosteric modulator of the cannabinoid CB1 receptor. Br J Pharmacol. 2015;172(20):4790–805.

30. Kulig K. Interpretation of workplace tests for cannabinoids. J Med Toxicol. 2017;13(1):106–10.

31. Huestis MA, Barnes A, Smith ML. Estimating the time of last cannabis use from plasma delta9-tetrahydrocannabinol and 11-nor-9-carboxy-delta9-tetrahydrocannabinol concentrations. Clin Chem. 2005;51(12):2289–95.

32. Huestis MA, Henningfield JE, Cone EJ. Blood cannabinoids. II. Models for the prediction of time of marijuana exposure from plasma concentrations of delta 9-tetrahydrocannabinol (THC) and 11-nor-9-carboxy-delta 9-tetrahydrocannabinol (THCCOOH). J Anal Toxicol. 1992;16(5):283–90.

33. Terhune KW et al. *The incidence and role of drugs in fatally injured drivers.* 1992(NTIS-PB94151768).

34. Narang S, et al. Efficacy of dronabinol as an adjuvant treatment for chronic pain patients on opioid therapy. J Pain. 2008;9(3):254–64.

Pediatric Considerations when Prescribing Cannabis

14

Shinya Ito and Ruud Verstegen

Abbreviations

[C]ss   Average serum concentration at steady state
CBD   cannabidiol (one of the major active components of cannabis)
CL   Clearance
Cmax   Peak concentration
F   Oral bioavailability
t1/2   Elimination half-life
THC   Δ9-Tetrahydrocannabinol (one of the major active components of cannabis)
Vd   Volume of distribution

Introduction

Except for neonates, children have a higher capacity per body weight than adults to eliminate drug, unless they have decreased liver and/or kidney function. This implies that they usually require a higher dose per body weight than adults to achieve the same

S. Ito (✉)
Division of Clinical Pharmacology & Toxicology, Hospital for Sick Children, Toronto, ON, Canada

Department of Paediatrics, University of Toronto, Toronto, ON, Canada
e-mail: shinya.ito@sickkids.ca

R. Verstegen
Division of Clinical Pharmacology & Toxicology, Hospital for Sick Children, Toronto, ON, Canada

Department of Paediatrics, University of Toronto, Toronto, ON, Canada

Division of Rheumatology, Hospital for Sick Children, Toronto, ON, Canada
e-mail: ruud.verstegen@sickkids.ca

© Springer Nature Switzerland AG 2022
R. Valani (ed.), *Cannabis Use in Medicine*,
https://doi.org/10.1007/978-3-031-12722-9_14

average serum concentration at steady state. Shorter elimination half-life of many drugs in children, including cannabinoids (a group of compounds which bind to endogenous cannabinoid receptors), reflect this pharmacokinetic property. Pediatric use of cannabinoids ranges from cannabis plant extracts in oil, such as THC (Δ9-Tetrahydrocannabinol) and CBD (cannabidiol), to synthetic forms of cannabinoids such as dronabinol (THC isomer) and nabilone (THC-related cannabinoid). While THC and THC-like cannabinoids are psychoactive, causing euphoria and altered mental state, CBD and its active metabolite (7-hydroxy CBD) do not show psychoactive property but have sedative and antiepileptic effects, which may be therapeutic.

Overall, pharmacokinetic and pharmacodynamic characteristics of THC and CBD are poorly characterized in children, but as indicated above, children are likely to have higher clearance of these compounds on a body weight basis. Drug-drug interactions involving THC and/or CBD may become clinically significant, and victim drugs include clobazam, topiramate, tacrolimus, brivaracetam, clopidogrel, warfarin and direct oral anticoagulant (dabigatran, apixaban, edoxaban and rivaroxaban). Inducers of CYP3A4 and P-glycoprotein (a drug transporter) such as rifampicin may decrease blood concentrations of THC, its active metabolite (11-hydroxy THC) and CBD. In contrast, CYP3A4 inhibitors including ketoconazole increases their blood concentrations. Although evidence of clinical significance of these interactions remains relatively low for some drugs, pediatricians should be aware of this potential effect.

Clinically Useful Principles of Pharmacokinetics, and its Characteristics in Children

- For any compound, including drugs, a constant continuous dosage over time (i.e., a consistent dose/time), will eventually reach a state of equilibrium known as a steady state.
  - At steady state, the amount of drug eliminated per time is equal to the amount administered to the system (dose/time).
  - At steady state, a system maintains a consistent average serum concentration of the drug, which is called [C]ss (See Fig. 14.1).
- Clearance (CL), also called plasma clearance or total body clearance, is the plasma volume that has all drug removed per time unit (e.g., mL/min, L/H).
- CL relates the average plasma concentrations of drug at steady state [C]ss to a dose given per time (e.g., 1 mg/day, 10 microgram/min, etc. See Fig. 14.1).
  - At a given dose/time, the higher CL, the lower [C]ss; and the lower CL, the higher [C]ss.
  - At a given CL, the higher dose per time, the higher [C]ss.
- The elimination half-life (t1/2) represents the time it takes to remove half of the drug amount from the system. This is determined by a ratio between CL and volume of distribution (Vd).
  - The higher CL, the shorter t1/2; and the lower CL, the longer t1/2.

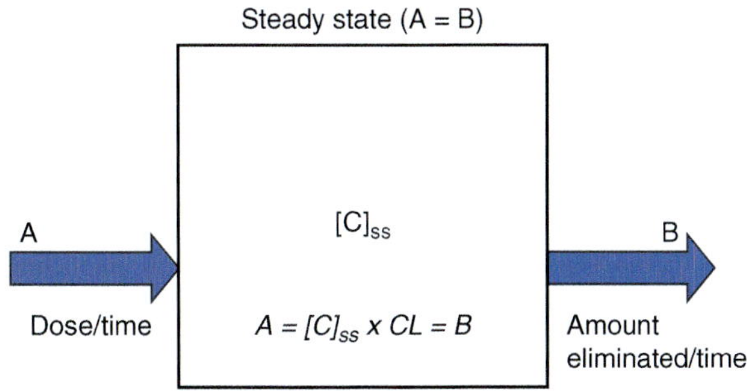

Fig. 14.1 At steady state, the dosing rate (A) equals the amount eliminated from the system per time (B), and CL relates [C]

- t1/2 is NOT dependent on the dose given (except when the elimination system is saturated).
- The larger Vd, the longer t1/2.
- It takes about 4 × t1/2 from the first dose to reach a steady state at a given dose and dosing interval.
- CL per body weight of most drugs is higher in young children (except neonates) than older children and adults.
  - Liver and kidney sizes per body weight are larger in young children compared to adults.
  - Once enzyme expression/function per liver tissue (or nephron per kidney tissue) is matured reaching an adult level, the enzyme amount (or nephron) per body weight is highest in young children.

    Therefore, to achieve the same [C]ss, young children often require higher dose per body weight than older children and adults.
  - Neonates have immature drug elimination systems.

    By 12–24 months of age, with some exceptions, the activity of most drug eliminating systems reaches a mature level per organ weight/volume.

    This leads to higher CL per body weight than older children and adults.

Pharmacological Properties of THC and CBD (See Chap. 3 on The Pharmacology of Cannabis)

- There are two main endogenous cannabinoid receptors in the body, namely CB_1 and CB_2 receptors:
  - The CB_1 receptor is highly expressed in neuronal tissue, both at the pre- and post- synaptic neurons.
  - The CB_2 receptor is located mainly in immune system.
- THC (Δ9-Tetrahydrocannabinol), dronabinol (a synthesized isoform of THC) and nabilone (a synthetic cannabinoid) have psychoactive activity by binding to the endogenous cannabinoid receptors.

- 11-OH-THC (11-hydroxy THC) has psychoactive properties as well.
  - This compound is a metabolite of THC and dronabinol.
- The metabolites of nabilone are not specified.
- CBD (cannabidiol).
  - CBD does *not* strongly bind to the cannabinoid receptors, and results in limited psychotropic effects (e.g., altering mental state).
  - However, it shows therapeutic function including antiepileptic, sedating and antiemetic effects.
- 7-OH-CBD (7-hydroxy CBD).
  - This compound is a major metabolite of CBD with similar activity and serum concentrations.

Elimination Pathways of THC and CBD [1–3]

- THC is eliminated through the liver by metabolism
  - THC is metabolized by a drug metabolizing enzyme (CYP2C9) to a psychoactive metabolite 11-OH-THC [4].
  - 11-OH-THC is metabolized by UGT1A9, UGT2B7 and CYP3A4 to inactive metabolites.
- CBD is eliminated through the liver by metabolism
  - CBD is metabolized by CYP2C19 and CYP3A4 [5].
  - 7-OH-CBD is probably glucuronidated and/or sulphated.

Known and Potentially Interacting Drug Elimination Pathways of Medical Cannabinoids

- THC may inhibit CYP2C9 [4].
- CBD may interfere with metabolism/transport of other drugs at P-glycoprotein (drug transporter), CYP2C8/9, CYP2C19 and CYP3A4/5 [6, 7].

Pharmacokinetic Parameters of Oral Medical Cannabinoids in Adults [8–11]

- THC, dronabinol, nabilone
  - Oral bioavailability (F)
    THC: 6–7% (range: 2–14%); dronabinol: 10–20%; nabilone: not reported. This low and variable bioavailability implies substantial inter-individual variations in systemically available drug after oral administration.
  - Cmax
    0.6–1.8 ng/mL (at a repeated daily dose of about 15 mg THC in hemp oil divided into 3 doses for 5 days, or dronabinol 7.5 mg divided into 3 doses for 5 days). Nabilone Cmax is 2 ng/mL.

- Time to reach Cmax:
  1 – 2 h post-dose (as late as 6 h).
- Plasma protein binding:
  95–99% to lipoproteins.
  Relatively high protein binding, suggesting that relatively small alterations in protein binding may lead to significant changes in unbound fraction. However, no drug interaction has been reported at protein binding.
- CL:
  0.2–1 L/min (intravenous)
  This highly variable CL is one of the reasons for substantial individual variations in achieved plasma concentrations.
- CL/F (oral):
  ~20 L/min.
  Coupled with variations in bioavailability (F), substantial individual variations exist.
- Vd:
  THC: 1–3 L/kg.
- Vd/F:
  THC: >10 L/kg (estimated from Vd and F)
  Dronabinol: 10 L/kg.
  Nabilone: 12.5 L/kg.
- t1/2:
  25–36 h (terminal phase).
- CBD [12]
  - Oral bioavailability (F):
    Around 10% (inhalation: 30%).
    Apparently increases at post-prandial dosing.
  - Cmax:
    5–15 ng/mL (at a single dose of 10–20 mg).
  - Time to reach Cmax:
    1 – 2 h (as late as 6 h).
  - Plasma protein binding:
    90%.
    Relatively high protein binding, suggesting that relatively small alterations in protein binding may potentially lead to significant changes in plasma concentrations of unbound CBD.
  - CL:
    1.2 L/min (intravenous).
    This is relatively high clearance for a hepatically eliminated compound.
  - CL/F:
    3–12 L/min (oral): about 40–160 mL/kg/min.
    Coupled with variations in bioavailability (F), substantial individual variations exist.
  - Vd:
    2500 L (or roughly 35 L/kg after intravenous administration).

- Vd/F:
    28,000 L (or roughly 400 L/kg).
    Extremely large volume of distribution.
- t1/2:
    CBD shows multi-phase elimination.
    Depending on the administration route and measured elimination phases, it ranges from 1–2 h (a single nebulizer/oral dose) to 5 days (reflecting a terminal elimination phase in chronic oral administration).

Pharmacokinetic Parameters of Oral Medical Cannabinoids in Children

- The pharmacokinetics of dronabinol and nabilone have not been reported for pediatric patients.
- THC
    - Oral bioavailability (F):
        Not reported.
    - Cmax:
        0.8–3.6 ng/mL (at a dose of 0.02–1.6 mg/kg).
    - Time to reach Cmax:
        2.7 h (range: 1–7 h).
    - CL or CL/F:
        Not reported.
        On a body weight basis, it is likely to be higher in children than adult values.
    - Vd or Vd/F:
        Not reported.
    - t1/2:
        4 h (range: 0.9–8.1 h)*
    *: This is not a terminal-phase half-life (see adult values above).
- CBD
    - Oral bioavailability (F):
        Not reported.
    - Cmax:
        120 ng/mL at steady state at Day 10 (5 mg/kg BID).
    - Time to reach Cmax:
        2–4 h (range: 1–24 h).
    - Plasma protein binding:
        Not reported (likely to be similar to adults).
    - CL (intravenous)
        Not reported.
    - CL/F:
        200–250 mL/kg/min.
    - Vd or Vd/F:
        Not reported, but probably similar to adult values.

- t1/2:
    20–30 h.

Pharmacogenetic Considerations (See Chap. 6 on The Role of Genetics in the Use of Cannabis)

- THC and dronabinol
    - THC and dronabinol are substrates of the CYP2C9 drug metabolizing enzyme [4].
    - A poor metabolizer variant of CYP2C9 tends to achieve 2–three-fold higher plasma concentrations.
    - The combined concentration of THC and its psychoactive metabolite 11-OH-THC is two-fold higher compared to normal metabolizers.
- Nabilone
    - No data available.
- CBD
    - CBD is a substrate of a drug metabolizing enzyme CYP2C19 [5].
    - CYP2C19 variant status may change CBD concentrations in plasma, but evidence is scarce.

Adverse Effects

- THC, dronabinol, nabilone
    - *Central nervous system*: somnolence, behavioural and mood changes, anxiety, sleep disturbance, impaired cognitive function, movement disorder, drooling, delirium, psychosis, seizure, visual disturbance, mydriasis, cerebrovascular accident.
    - *Cardiovascular*: tachycardia, hypotension, hypertension, flushing.
    - *Gastrointestinal*: anorexia (weight loss), xerostomia, nausea, vomiting (including cannabinoid hyperemesis syndrome), diarrhea, constipation, increased liver enzymes.
    - *Other*: Skin rash, anemia, hypersensitivity reaction, increased creatinine, infections (pneumoniae), anhidrosis, diaphoresis, increased/decreased micturition, musculoskeletal pain, eye irritation, tinnitus, cough, dyspnea, nasal congestion, epistaxis.
- CBD
    - *Central nervous system*: somnolence, behavioural changes, sleep disturbance, movement disorder, drooling.
    - *Gastrointestinal*: anorexia (weight loss), diarrhea, increased liver enzymes.
    - *Other*: Skin rash, anemia, hypersensitivity reaction, increased creatinine, infections.

Cannabis-drug Interactions [13]

- Depending on the compound/ drug, they can either induce or inhibit the metabolism of cannabinoids. Therefore, the clinician must be aware of these potential interactions.
  - CYP3A4 inhibitors. [6]
    These compounds can increase THC, 11-OH-THC and CBD serum concentrations. They may also have the same effects on dronabinol.
    Examples include:
    - Ketoconazole.
    - Macrolide antibiotics.
    - Rifampicin (It is also a P-glycoprotein inducer).
  - CYP3A4 inducers. [14]
    Inducers can decrease THC, 11-OH-THC and CBD serum concentrations. They may also have the same effects on dronabinol.
    Examples include:
    - Rifampicin (It is also a P-glycoprotein inducer).
  - CYP2C9 inhibitors. [4]
    These compounds will increase the levels of THC and dronabinol.
    Examples include:
    - Fluconazole.
    - Metronidazole.
    - Miconazole.
    - Voriconazole.
    - Phenytoin.
    - Co-trimoxazole.
    - Valproic acid [15].
- Other specific drug interactions to be aware of are:
  - Clobazam. (See Chap. 9 on Neurological diseases and cannabinoid treatment) [15–17]
    CBD can increase serum concentrations of clobazam and its active metabolite, nor-clobazam.
    Clobazam may increase serum concentrations of 7-OH-CBD (an active metabolite of CBD).
  - Topiramate. [18]
    CBD use may be associated with mildly elevated serum concentrations of topiramate.
    The clinical significance of this is unknown.
  - Eslicarbazepine, zonisamide and rufinamide.
    CBD use may be associated with increased serum concentrations of these drugs.
    The clinical significance of this is unknown.
  - Brivaracetam. [19]
    CBD may increase serum concentrations of brivaracetam by 100% or more.

The clinical significance is not clear, but somnolence was reported in some patients.

- Warfarin. [20]

  CBD or cannabis smoking have been shown to increase warfarin effects probably by inhibiting warfarin metabolism via CYP2C9.

  Warfarin dose reduction may be necessary and can be monitored with INR checks.

- Direct Oral Anti-Coagulant (DOAC, e.g., dabigatran, apixaban, edoxaban and rivaroxaban) [20].

  DOACs are P-glycoprotein substrates, and CBD-mediated inhibition of P-glycoprotein may increase their systemic exposures.

  Rivaroxaban and apixaban are CYP3A4 substrates, and CBD-mediated CYP3A4 inhibition may increase their systemic exposure.

  Rivaroxaban exposure levels were shown to increase by 30–160% in the presence of P-glycoprotein inhibitors.

- Clopidogrel. [20]

  CBD inhibits CYP2C19, which is an enzyme that activates Clopidogrel. CYP2C19 inhibition and/or reduced function is a risk factor for clopidogrel therapeutic failure [5].

- Tacrolimus. [21]

  CBD may significantly increase tacrolimus blood concentrations. Close therapeutic drug monitoring is recommended.

Pearls for Pediatric Dosing of Medical Cannabis

- There is a variety of cannabinoids available for medical use (medical cannabinoids):
  - THC (extraction in oil)
  - Synthetic THC (dronabinol)
  - Synthetic cannabinoid (nabilone)
  - CBD (extraction in oil)—less psychotropic compared to other compounds
- Pharmacokinetics for pediatric patients:
  - Approximately 10% of a medical cannabinoid dose is absorbed and peak levels are reached within 1–2 h.
  - Medical cannabinoids distribute primarily to the central nerve system and fat tissue (thus resulting in a high distribution volume).
  - Each compound has specific metabolic pathways; individual differences in these pathways (pharmacogenetics) may explain variations of their clinical efficacy and drug-drug interactions [4].
  - The half-life of medical cannabinoids is around 30 h.
- Pay attention to adverse effects, especially in children.
  - Adverse effects are more likely with THC/dronabinol/nabilone, compared to CBD.

- Central nervous system (altered mental status and behaviour), cardiovascular (tachycardia, hypo/hypertension) and gastrointestinal effects (anorexia, vomiting, diarrhea) are most common.
- Drug-drug interactions with the following medications may be present:
  - THC/dronabinol: antifungals (azoles), and rifampicin.
  - CBD: clobazam, topiramate, eslicarbazepine, zonisamide, rufinamide, brivaracetam, warfarin, DOACs, clopidogrel, and tacrolimus.

References

1. Lucas CJ, Galettis P, Schneider J. The pharmacokinetics and the pharmacodynamics of cannabinoids. Br J Clin Pharmacol. 2018;84(11):2477–82.
2. Mouhamed Y, et al. Therapeutic potential of medicinal marijuana: an educational primer for health care professionals. Drug Healthc Patient Saf. 2018;10:45–66.
3. Alsherbiny MA, Li CG. Medicinal Cannabis-potential drug interactions. Medicines (Basel). 2018;6:1.
4. Sachse-Seeboth C, et al. Interindividual variation in the pharmacokinetics of Delta9-tetrahydrocannabinol as related to genetic polymorphisms in CYP2C9. Clin Pharmacol Ther. 2009;85(3):273–6.
5. Jiang R, et al. Cannabidiol is a potent inhibitor of the catalytic activity of cytochrome P450 2C19. Drug Metab Pharmacokinet. 2013;28(4):332–8.
6. Yamaori S, et al. Potent inhibition of human cytochrome P450 3A isoforms by cannabidiol: role of phenolic hydroxyl groups in the resorcinol moiety. Life Sci. 2011;88(15–16):730–6.
7. Zhu HJ, et al. Characterization of P-glycoprotein inhibition by major cannabinoids from marijuana. J Pharmacol Exp Ther. 2006;317(2):850–7.
8. Millar SA, et al. A Systematic review on the pharmacokinetics of cannabidiol in humans. Front Pharmacol. 2018;9:1365.
9. Grotenhermen F. Pharmacokinetics and pharmacodynamics of cannabinoids. Clin Pharmacokinet. 2003;42(4):327–60.
10. Taylor L, et al. A phase I, randomized, double-blind, placebo-controlled, single ascending dose, multiple dose, and food effect trial of the safety, tolerability and pharmacokinetics of highly purified cannabidiol in healthy subjects. CNS Drugs. 2018;32(11):1053–67.
11. Tayo B, et al. A phase I, open-label, parallel-group, single-dose trial of the pharmacokinetics, safety, and tolerability of cannabidiol in subjects with mild to severe renal impairment. Clin Pharmacokinet. 2020;59(6):747–55.
12. Ohlsson A, et al. Single-dose kinetics of deuterium-labelled cannabidiol in man after smoking and intravenous administration. Biomed Environ Mass Spectrom. 1986;13(2):77–83.
13. Qian Y, Gurley BJ, Markowitz JS. The potential for pharmacokinetic interactions between cannabis products and conventional medications. J Clin Psychopharmacol. 2019;39(5):462–71.
14. Stott C, et al. A Phase I, open-label, randomized, crossover study in three parallel groups to evaluate the effect of Rifampicin, Ketoconazole, and Omeprazole on the pharmacokinetics of THC/CBD oromucosal spray in healthy volunteers. Springerplus. 2013;2(1):236.
15. Gaston TE, et al. Interactions between cannabidiol and commonly used antiepileptic drugs. Epilepsia. 2017;58(9):1586–92.
16. Geffrey AL, et al. Drug-drug interaction between clobazam and cannabidiol in children with refractory epilepsy. Epilepsia. 2015;56(8):1246–51.

17. Morrison G, et al. A Phase 1, Open-Label, Pharmacokinetic Trial to Investigate Possible Drug-Drug Interactions Between Clobazam, Stiripentol, or Valproate and Cannabidiol in Healthy Subjects. Clin Pharmacol Drug Dev. 2019;8(8):1009–31.
18. Luna-Tortós C, et al. The antiepileptic drug topiramate is a substrate for human P-glycoprotein but not multidrug resistance proteins. Pharm Res. 2009;26(11):2464–70.
19. Klotz KA, et al. Effects of cannabidiol on brivaracetam plasma levels. Epilepsia. 2019;60(7):e74–7.
20. Greger J, et al. A review of cannabis and interactions with anticoagulant and antiplatelet agents. J Clin Pharmacol. 2020;60(4):432–8.
21. Leino AD, et al. Evidence of a clinically significant drug-drug interaction between cannabidiol and tacrolimus. Am J Transplant. 2019;19(10):2944–8.

Cannabis Use in the Pregnant Patient

15

Prabhpreet Hundal and Simina Luca

Introduction

The National Survey on Drug Use and Health in the United States found that between 2007 and 2012, more than 1 in 10 women of child-bearing age used marijuana in the prior 12 months [1]. With cannabis becoming more permissive, it is not surprising that consumption has increased exponentially over the last few years. While the prevalence of use across North America is still low at 2–4%, there has been a 62% increase from 2002 to 2014 among pregnant women using cannabis with 11–16.2% using it on a daily basis [2–4]. Therefore, it is important for clinicians to understand and appreciate the health impacts of cannabis on pregnant women and fetuses.

Effects of Cannabis in Pregnancy and Breastfeeding

- $\triangle$9-Tetrahydrocannabinol (THC) is the main psychoactive constituent of cannabis, which can be consumed in many ways: smoking, vaping, eating, and dabbing (breathing in hot vapours released by heating cannabis concentrates). See Chap. 3 on the pharmacology of cannabis.
  - When compared to smoking, vaping leads to a higher concentration of cannabinoids in blood and oral fluids [5].

P. Hundal (✉)
Department of Obstetrics and Gynecology, University of Alberta, Edmonton, AB, Canada
e-mail: phundal@ualberta.ca

S. Luca
William Osler Health System, Brampton, ON, Canada

McMaster University, Hamilton, ON, Canada

University of Toronto, Toronto, ON, Canada

© Springer Nature Switzerland AG 2022
R. Valani (ed.), *Cannabis Use in Medicine*,
https://doi.org/10.1007/978-3-031-12722-9_15

- THC concentrations peak faster and are seen in higher concentration when inhaled compared to oral route [6].
- In pregnant patients who consume cannabis, THC crosses the placenta and enters the fetal bloodstream. The concentration of THC depends on the route of administration [7].
  - Inhaled THC is associated with a three-times higher fetal concentration compared to ingested THC [8].
- THC is also excreted in breast milk. One study found that 63% of breast milk samples had detectable levels of THC [9].
  - While the exact concentration of THC varied, higher concentrations were found in those who:
    Used it more frequently and on a daily basis.
    Used it via an inhaled route versus ingested.
  - THC can be present in breast milk for up to 6 days after its last use.
- Cannabis affects the fetal endocannabinoid signaling system, which forms early in gestation. This system is responsible for the development of the brain, neuronal connectivity, and glial cell differentiation through neuromodulation of various central neurotransmitter systems [10].
  - THC exposure to the fetus leads to exogenous cannabinoids, which can impair fetal growth and neurodevelopment through its adverse effects on the maturation of various neurotransmitter systems.
  - In addition, the cannabinoid system plays a role in immunomodulation. It is not fully understood what the implications of external cannabinoids on the fetus immune system.
- Given the lipophilicity of the THC molecule, it selectively accumulates in fat-rich organs, such as the brain [11].
  - As a result, cannabinoids are likely to affect brain growth and development when it occurs rapidly, which is during the first two years of life. This time period coincides with the time period when children are also breastfed [11].

Indications for Cannabis in Pregnancy

- There are numerous potential therapeutic uses of cannabis that have been recognized, such as for chemotherapy-induced nausea and vomiting [10].
  - However, cannabis is not indicated for the management of any medical conditions in pregnancy and is particularly not recommended for pregnancy-related nausea or emesis.
- Women with depression are three times more likely to use cannabis during pregnancy [12].
- The American College of Obstetricians and Gynecologists does not recommend the use of either medical or non-medical cannabis during pregnancy (See Section

on Recommendations from Professional Societies) (The American College of Obstetricians and Gynecologists).

Cannabis Use for Pregnancy-Related Nausea and Emesis

- Though not medically indicated, some women use cannabis as an antiemetic during pregnancy [13, 14].
- Studies have shown that women experiencing severe nausea related to pregnancy are more likely to use cannabis (3.7% vs 2.3%, respectively) [15].
  - Interestingly, another study found that 92% of women who used cannabis to treat pregnancy-related nausea or vomiting rated it to be effective or extremely effective [16].
- Given the potential implications of using this drug during the first trimester, the American Academy of Pediatrics recommends that women who use cannabis to treat a medical condition should be advised to use alternative treatments that have better safety-profiles in pregnancy [11].

The Impacts of Cannabis on Pregnancy and the Fetus (See Fig. 15.1)

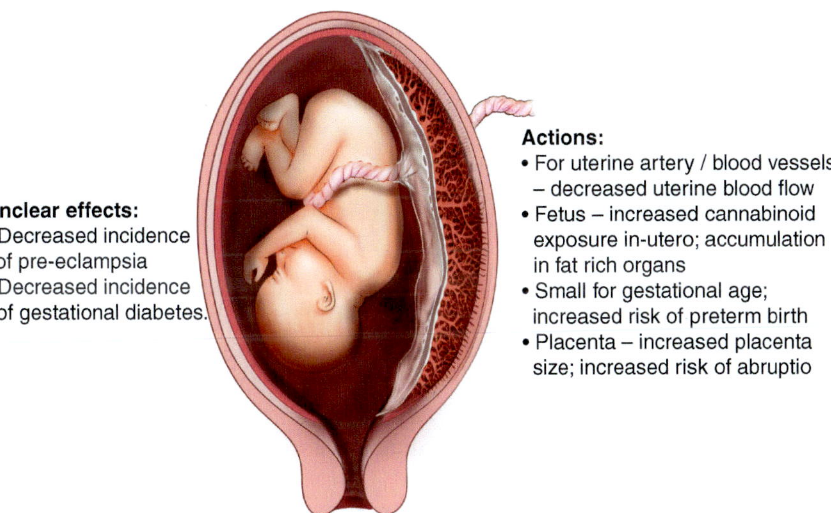

Fig. 15.1 The effects of cannabis use during pregnancy on the fetus>

On the Placenta

- Prenatal cannabis exposure has been linked to the development of the placenta and the fetus [17].
  - Cannabis use in pregnancy has been associated with an increased pulsatility and resistance index of the uterine artery, with resulting potential decrease in uterine blood flow [18].
  - There is an association with larger placental size in women who used cannabis during pregnancy. This may be compensatory change due to chronic hypoxia to the placenta and fetus [19].
- One study found that the rate of placental abruption was higher in women reporting cannabis use [19].

On Preterm Birth

- While there is conflicting evidence on this topic, a recent study found that cannabis use during pregnancy was significantly associated with preterm birth [20].
  - Additionally, they found a significantly increased risk of:
    Small for gestational age babies,
    The need for neonatal intensive care, and
    A lower 5-min Apgar score.
- In patients who used tobacco as well as cannabis, the risk of preterm birth was even higher [21].
- According to the SCOPE trial, the timing of cannabis use (after 20 weeks gestation) was associated with a five-fold increased risk of preterm delivery [22].
- A French study found that cannabis consumption was one of the main factors for preterm birth, along with: [23]
  - Prior history of preterm birth,
  - Primiparity,
  - Low body mass index,
  - Low educational level, and
  - Limited prenatal care.

On Birth Weight

- There is significant controversy regarding the association between cannabis use and low birth weight.
- Several studies have found an association between cannabis use and small for gestational age babies [24].
  - There is a 30–50% likelihood of having low birth weight with cannabis use, with an odds ratio of 2.72 [20, 25–27].
  - Studies have found a decrease in weight of 100–565 g in the new baby [28, 29].

- Unfortunately, the clinical significance of this is unknown, and other studies have not found any association between cannabis use in the third trimester and changes in birthweight [30].

On Neonatal Intensive Care Unit (NICU) Admission

- There is conflicting evidence on the rate of NICU admission for mothers who use cannabis, with some studies reporting a higher rate of admission while others found no association [4, 25, 26, 31].
- For cannabis use in the third trimester, one study noted a relationship between babies requiring medication for neonatal abstinence syndrome and a longer stay in hospital with in-utero cannabis exposure. However, this relationship was not statistically significant [30].

On Stillbirth

- There is an association between cannabis use and increased risk of neonatal death [31].
  - An increased risk of stillbirth with cannabis use in pregnancy, with an odds ratio of 2.34, was also demonstrated by the Stillbirth Collaborative Research Network.

On Congenital Anomalies

- It was previously believed that there were no associated congenital anomalies secondary to cannabis use in pregnancy. With the substantially higher levels of THC in contemporary cannabis consumption, a study published in 2019 looked at the rates of congenital defects in Colorado. It found that the trend of rising congenital defects closely paralleled the rise in cannabis use in this state, in the context of static or falling levels of other drug use [32].
  - Congenital defects included atrial septal defect, patent ductus arteriosus, ventricular septal defect, Down's syndrome, spina bifida and microcephalus among others.
- This study suggests an association only, so further research is certainly required before making any definitive statements in this respect.

On Other Pregnancy Related Conditions

- Interestingly, a study found that cannabis use decreased the risk of pre-eclampsia and gestational diabetes, the mechanisms of which are not fully understood [20].

On the Child

- Two cohort studies have investigated the long-term impact of in-utero cannabis exposure: the Ottawa Prenatal Prospective Study and the Maternal Health Practices and Child Development Study [7, 8].
    - These studies found an association between impaired neurodevelopment among children exposed to cannabis in utero. After the age of 3, their cognitive functions, such as memory and verbal skills, were adversely affected.
    - In children aged 9–12, impairment of integrative tasks and analytical processing were noted. By the age of 14, poorer school performance was noted in children exposed to in utero cannabis.
- Individuals exposed to cannabis in utero may be more likely to use cannabis during young adulthood [33].

On Breastfeeding

- Increased depressive symptoms and a shorter duration of breastfeeding were both associated with the use of cannabis in the postpartum period [3].

Recommendations from Professional Societies

American College of Obstetricians and Gynecologists (ACOG) (Committee on Obstetric Practice)

- Women who report cannabis use should be counselled about the potential health consequences in pregnancy and encouraged to discontinue its use before and during pregnancy.
- Given the lack of sufficient data to evaluate the impact of cannabis use during breastfeeding, its use should be discouraged.
- Women who are using cannabis for medicinal purposes should be advised to discontinue its use in favour of therapies with better safety data during pregnancy.

Society of Obstetricians and Gynecologists of Canada (SOGC) [34]

- Women who occasionally or regularly use cannabis should be advised to abstain from or reduce cannabis use during pregnancy and while breastfeeding.

Royal Australian and New Zealand College of Obstetricians and Gynecologists (RANZCOG) (Women's Health Committee)

- Women who are pregnant or those planning pregnancy should be advised to discontinue cannabis use, given the findings of neurodevelopment delay in neonates and children of women who used cannabis in pregnancy and breastfeeding.

American Academy of Breastfeeding Medicine (ABM) [35]

- Current data cannot be used to recommend women stop breastfeeding, however, the ABM cautions against cannabis use while breastfeeding.
- Women who report cannabis use while breastfeeding should be advised to reduce or avoid its use, in order to minimize the long-term neurobehavioural effects associated with cannabis. Infants should not be exposed to cannabis or its smoke.

American Academy of Pediatrics [36]

- Women who are pregnant or are considering pregnancy should be counselled about the current concerns related to the impact of cannabis on pregnancy, the fetus, and child development, even though there is a lack of definitive research.
- Cannabis should not be used during pregnancy.
- Women who are using cannabis for medical conditions or to treat nausea and vomiting during pregnancy should be advised to speak with their healthcare provider about alternative treatments with safety data during pregnancy.
- Given the lack of data about adverse effects, women should be counselled to abstain from cannabis use while breastfeeding.
- Infant exposure to smoke from cannabis should be minimized.

Other Implications

- It is important to screen every pregnant patient for substance abuse, and educate them on its effects on pregnancy and their baby. However, it is also crucial to do so in a non- judgmental way that promotes open discussion and sometimes, harm reduction [37].

Key Summary Points

- The rates of cannabis use in pregnancy are increasing.
- THC crosses the placenta and enters the fetal bloodstream, selectively depositing in fat- rich organs such as the brain. It impacts the fetal endocannabinoid signaling system and can impair fetal growth and neurodevelopment. It may also impact the immune system of the fetus.
- THC can be detected in breastmilk, and new mothers need to be informed of the implications of this.
- Currently, there are no indications for cannabis use in pregnancy. Women may be using cannabis as an antiemetic during pregnancy. However, they should be advised to use alternative treatments with better safety-profiles in pregnancy.
- While there are mixed reviews, most studies suggest an association between cannabis use and increased risk of preterm birth and low birth weight.

- In-utero cannabis exposure is associated with increased long-term neurodevelopmental issues.
- Current medical professional organizations recommend abstaining from cannabis use during pregnancy and the postpartum period.

References

1. Ko JY, et al. Prevalence and patterns of marijuana use among pregnant and nonpregnant women of reproductive age. Am J Obstet Gynecol. 2015;213(2):201.e1–201.e10.
2. Brown QL, et al. Trends in marijuana use among pregnant and nonpregnant reproductive-aged women, 2002–2014. JAMA. 2017;317(2):207–9.
3. Ko JY, et al. Marijuana use during and after pregnancy and association of prenatal use on birth outcomes: a population-based study. Drug Alcohol Depend. 2018;187:72–8.
4. Corsi DJ, et al. Trends and correlates of cannabis use in pregnancy: a population-based study in Ontario, Canada from 2012 to 2017. Can J Public Health. 2019;110(1):76–84.
5. Spindle TR, et al. Acute pharmacokinetic profile of smoked and vaporized cannabis in human blood and Oral fluid. J Anal Toxicol. 2019;43(4):233–58.
6. Newmeyer MN, et al. Free and glucuronide whole blood Cannabinoids' pharmacokinetics after controlled smoked, vaporized, and Oral Cannabis Administration in Frequent and Occasional Cannabis Users: identification of recent cannabis intake. Clin Chem. 2016;62(12):1579–92.
7. Jansson LM, Jordan CJ, Velez ML. Perinatal marijuana use and the developing child. JAMA. 2018;320(6):545–6.
8. Foeller ME, Lyell DJ. Marijuana use in pregnancy: concerns in an evolving era. J Midwifery Womens Health. 2017;62(3):363–7.
9. Bertrand KA, et al. Marijuana use by breastfeeding mothers and cannabinoid concentrations in breast Milk. Pediatrics. 2018;142:3.
10. Government of Canada. *Information for health care professionals. Cannabis (marihuana, marijuana) and the cannabinoids.* 2018.; https://www.canada.ca/content/dam/hc-sc/documents/services/drugs-medication/cannabis/information-medical-practitioners/information-health-care-professionals-cannabis-cannabinoids-eng.pdf.
11. Ryan SA, Ammerman SD, O'Connor ME. Marijuana use during pregnancy and breastfeeding: implications for neonatal and childhood outcomes. Pediatrics. 2018;142:3.
12. Goodwin RD, et al. Cannabis use during pregnancy in the United States: the role of depression. Drug Alcohol Depend. 2020;210:107881.
13. Volkow ND, Compton WM, Wargo EM. The risks of marijuana use during pregnancy. JAMA. 2017;317(2):129–30.
14. Young-Wolff KC, et al. Trends in marijuana use among pregnant women with and without nausea and vomiting in pregnancy, 2009–2016. Drug Alcohol Depend. 2019;196:66–70.
15. Roberson EK, Patrick WK, Hurwitz EL. Marijuana use and maternal experiences of severe nausea during pregnancy in Hawai'i. Hawaii J Med Public Health. 2014;73(9):283–7.
16. Westfall RE, et al. Reprint of: survey of medicinal cannabis use among childbearing women: patterns of its use in pregnancy and retroactive self-assessment of its efficacy against 'morning sickness'. Complement Ther Clin Pract. 2009;15(4):242–6.
17. Nashed MG, Hardy DB, Laviolette SR. Prenatal cannabinoid exposure: emerging evidence of physiological and neuropsychiatric abnormalities. Front Psych. 2020;11:624275.
18. El Marroun H, et al. A prospective study on intrauterine cannabis exposure and fetal blood flow. Early Hum Dev. 2010;86(4):231–6.
19. Carter RC, et al. Alcohol, methamphetamine, and marijuana exposure have distinct effects on the human placenta. Alcohol Clin Exp Res. 2016;40(4):753–64.

20. Corsi DJ, et al. Association between self-reported prenatal cannabis use and maternal, perinatal, and neonatal outcomes. JAMA. 2019;322(2):145–52.
21. Chabarria KC, et al. Marijuana use and its effects in pregnancy. Am J Obstet Gynecol. 2016;215(4):506.e1-7.
22. Leemaqz SY, et al. Maternal marijuana use has independent effects on risk for spontaneous preterm birth but not other common late pregnancy complications. Reprod Toxicol. 2016;62:77–86.
23. Prunet C, et al. Risk factors of preterm birth in France in 2010 and changes since 1995: results from the French National Perinatal Surveys. J Gynecol Obstet Hum Reprod. 2017;46(1):19–28.
24. Michalski CA, et al. Association between maternal cannabis use and birth outcomes: an observational study. BMC Pregnancy Childbirth. 2020;20(1):771.
25. El Marroun H, et al. Prenatal cannabis and tobacco exposure in relation to brain morphology: a prospective neuroimaging study in young children. Biol Psychiatry. 2016;79(12):971–9.
26. Crume TL, et al. Cannabis use during the perinatal period in a state with legalized recreational and medical marijuana: the association between maternal characteristics, breastfeeding patterns, and neonatal outcomes. J Pediatr. 2018;197:90–6.
27. Campbell EE, et al. Socioeconomic status and adverse birth outcomes: a population-based Canadian sample. J Biosoc Sci. 2018;50(1):102–13.
28. Gunn JK, et al. Prenatal exposure to cannabis and maternal and child health outcomes: a systematic review and meta-analysis. BMJ Open. 2016;6(4):e009986.
29. Brown SJ, et al. Use of cannabis during pregnancy and birth outcomes in an aboriginal birth cohort: a cross-sectional, population-based study. BMJ Open. 2016;6(2):e010286.
30. O'Connor AB, Kelly BK, O'Brien LM. Maternal and infant outcomes following third trimester exposure to marijuana in opioid dependent pregnant women maintained on buprenorphine. Drug Alcohol Depend. 2017;180:200–3.
31. Metz TD, et al. Maternal marijuana use, adverse pregnancy outcomes, and neonatal morbidity. Am J Obstet Gynecol. 2017;217(4):478.e1–8.
32. Reece AS, Hulse GK. Cannabis teratology explains current patterns of Coloradan congenital defects: the contribution of increased cannabinoid exposure to rising teratological trends. Clin Pediatr (Phila). 2019;58(10):1085–123.
33. Sonon KE, et al. Prenatal marijuana exposure predicts marijuana use in young adulthood. Neurotoxicol Teratol. 2015;47:10–5.
34. Ordean A, Wong S, Graves L, No. 349-substance use in pregnancy. J Obstet Gynaecol Can. 2017;39(10):922-937.e2.
35. Reece-Stremtan S, Marinelli KA. ABM clinical protocol #21: guidelines for breastfeeding and substance use or substance use disorder, revised 2015. Breastfeed Med. 2015;10(3):135–41.
36. Bergeria CL, Heil SH. Surveying lactation professionals regarding marijuana use and breastfeeding. Breastfeed Med. 2015;10(7):377–80.
37. Miller J. Ethical issues arising from marijuana use by nursing mothers in a changing legal and cultural context. HEC Forum. 2019;31(1):11–27.

© Springer Nature Switzerland AG 2022
R. Valani (ed.), *Cannabis Use in Medicine*,
https://doi.org/10.1007/978-3-031-12722-9